PRAISE FOR *GENIUS MAKERS*

"Unlike many of the books written about AI, you don't need a science or engineering degree to learn from and enjoy this one. Anyone with an enthusiastic curiosity about science, technology, and the future of human culture will find this clear-eyed, snappily written book both entertaining and valuable. You could even call it essential for any policymakers, politicians, police, lawyers, judges, and decision-makers who will be contending with the social forces unleashed by artificial intelligence. Which, soon, will mean all of them."

—*Los Angeles Times*

"[An] engaging new book . . . [Metz's] straightforward writing perfectly translates industry jargon for technologically un-savvy readers (like me) who might be unfamiliar with what it means for a machine to engage in 'deep learning' or master tasks through its own experiences."

—*Christian Science Monitor*

"Carving a narrative out of a complex and ever-changing cast of characters . . . the book is filled with enlightening anecdotes that add texture and drama to the story. *Genius Makers* opens with Geoffrey Hinton, the Brit turned Canadian who is widely recognized as having played the most critical role in developing deep learning, the branch of AI that is changing the world today."

—*The Washington Post*

"A ringside seat at what may turn out to be *the* pivotal episode in human history. . . . Metz has a breezy style that is easy and fun to read. . . . undeniably charming."

—*Forbes*

"Colorful and readable . . . draws on extensive access and meticulous research."

—*Financial Times*

"Valuably suggests a framework for the right questions to ask now about AI and its use. *Genius Makers* is about the people who have built the AI world. . . ."

—James Fallows, *New York Times Book Review*

"The first book to chronicle the rise of savant-like artificial intelligence, and the last we'll ever need. . . . A ripping good read."

—William Softky, *Fair Observer*

"An informative, enjoyable work . . . With vivid detail, Metz has crafted an accessible narrative that will keep readers turning the pages."

—*Library Journal* (starred review)

"A must-read, fully up-to-date report on the holy grail of computing."

—*Kirkus Reviews* (starred review)

"With well-crafted storytelling and extensive research, Metz captures the thrill and promise of technological innovation."

—*Booklist*

"Written by an expert who has exclusive access to each of these companies—and others who are working in this field—this is a rich, character-driven narrative that captures an extraordinary moment in the history of technology."

—*Irish Tech News*

"In *Genius Makers*, Cade Metz delivers the definitive take on how AI technology came to be and what its arrival will mean for us humans. The book relies on tireless reporting and delightful writing to bring to life one of the most surprising and important stories of our time. If you want to read one book to understand AI, this is the one."

—Ashlee Vance, *New York Times* bestselling author of *Elon Musk*

"This colorful page-turner puts artificial intelligence into a human perspective. Through the lives of Geoff Hinton and other major players, Metz explains this transformative technology, and makes the quest thrilling."

—Walter Isaacson, #1 *New York Times* bestselling author of *Leonardo da Vinci*, *Steve Jobs*, and *The Innovators*

"Cade Metz has produced an enthralling narrative of the advance of artificial intelligence. He describes the key personalities, the seminal meetings, and the crucial breakthroughs with his customary eye for detail, building them into a dramatic history of this era-defining technology."

—Kai-Fu Lee, author of *AI Superpowers*

"This is the inside story of how AI entered Google, Facebook, and the rest of high tech. It is also the story of how Silicon Valley and its megabucks infiltrated AI and changed its course. Chock-full of behind-the-scenes anecdotes and wry humor—we learn the true tale of the technology that is transforming humanity."

—Oren Etzioni, chief executive, Allen Institute for Artificial Intelligence

"One day soon, when computers are safely driving on our roads and speaking to us in complete sentences, we'll look back at Cade Metz's elegant, sweeping *Genius Makers* as their birth story—the Genesis for an age of sentient machines."

—Brad Stone, author of *The Everything Store*, *The Upstarts*, and *Amazon Unbound*

"*Genius Makers* is an enthralling, definitive modern history of artificial intelligence. Cade Metz's detailed narrative reveals the crucial decisions made by executives, developers, and investors—and foreshadows the disproportionately large effect they will have on our futures."

—Amy Webb, author of *The Big Nine*

GENIUS MAKERS

THE MAVERICKS WHO BROUGHT AI
TO GOOGLE, FACEBOOK, AND THE WORLD

CADE METZ

DUTTON

An imprint of Penguin Random House LLC
penguinrandomhouse.com

Previously published as a Dutton hardcover in March 2021

First Dutton trade paperback edition: February 2022

THE LIBRARY OF CONGRESS HAS CATALOGED THE HARDCOVER EDITION OF THIS BOOK AS FOLLOWS:

Names: Metz, Cade, author.
Title: Genius makers: the mavericks who brought AI to
Google, Facebook, and the world / Cade Metz.
Description: New York: Dutton, an imprint of Penguin Random House LLC, [2021] |
Includes bibliographical references and index.
Identifiers: LCCN 2020041075 (print) | LCCN 2020041076 (ebook) |
ISBN 9781524742676 (hardcover) | ISBN 9781524742683 (ebook)
Subjects: LCSH: Artificial intelligence—Industrial applications—History. |
Intelligent personal assistants (computer software)—History. |
Machine learning—Research—History. | Computer scientists—Biography.
Classification: LCC TA347.A78 M48 2021 (print) |
LCC TA347.A78 (ebook) | DDC 006.3092/2—dc23
LC record available at https://lccn.loc.gov/2020041075
LC ebook record available at https://lccn.loc.gov/2020041076

Dutton trade paperback ISBN: 9781524742690

Printed in the United States of America
2nd Printing

In memory of Walt Metz,
who believed in truth, goodness, and beauty

It’s the best possible time to be alive, when almost everything you thought you knew is wrong.

—Tom Stoppard, *Arcadia*, Act One, Scene IV

When we have found all the mysteries and lost all the meaning, we will be alone, on an empty shore.

—Act Two, Scene VII

CONTENTS

PART TWO

WHO OWNS INTELLIGENCE?

PART THREE

TURMOIL

PART FOUR

HUMANS ARE UNDERRATED

PROLOGUE

THE MAN WHO DIDN'T SIT DOWN

DECEMBER 2012

By the time he stepped onto the bus in downtown Toronto, bound for Lake Tahoe, Geoff Hinton hadn't sat down for seven years. "I last sat down in 2005," he often said, "and it was a mistake." He first injured his back as a teenager, while lifting a space heater for his mother. As he reached his late fifties, he couldn't sit down without risking a slipped disk, and if it slipped, the pain could put him in bed for weeks. So he stopped sitting down. He used a standing desk inside his office at the University of Toronto. When eating meals, he put a small foam pad on the floor and knelt at the table, poised like a monk at the altar. He lay down when riding in cars, stretching across the back seat. And when traveling longer distances, he took the train. He couldn't fly, at least not with the commercial airlines, because they made him sit during takeoff and landing. "It got to the point where I thought I might be crippled—that I wouldn't be able to make it through the day—so I took it seriously," he says. "If you let it completely control your life, it doesn't give you any problems."

That fall, before lying down at the back of the bus for the trip to New York, taking the train all the way to Truckee, California, at the crest of the Sierra Nevadas, and then stretching across the back seat of a taxi for the thirty-minute drive up the mountain to Lake Tahoe, he created a new company. It included only two other people, both young graduate students in his lab at the university. It made no products. It had no plans to make a product. And its website offered nothing but a name, DNNresearch, which was even less appealing than the website. The sixty-four-year-old Hinton—who seemed so at home in academia, with his tousled gray hair, wool sweaters, and two-steps-ahead-of-you sense of humor—wasn't even sure he wanted to start a company until his two students talked him into it. But as he arrived in Lake Tahoe, one of the largest companies in China had already offered $12 million for his newborn start-up, and soon three other companies would join the bidding, including two of the largest in the United States.

He was headed for Harrah's and Harvey's, the two towering casinos at the foot of the ski mountains on the south side of the lake. Rising up over the Nevada pines, these twin slabs of glass, steel, and stone also served as convention centers, offering hundreds of hotel rooms, dozens of meeting spaces, and a wide variety of (second-rate) restaurants. That December, they hosted an annual gathering of computer scientists called NIPS. Short for Neural Information Processing Systems—a name that looked deep into the future of computing—NIPS was a conference dedicated to artificial intelligence. A London-born academic who had explored the frontiers of AI at universities in Britain, the United States, and Canada since the early 1970s, Hinton made the trip to NIPS nearly every year. But this was different. Although Chinese interest in his company was already locked in, he knew that others were interested, too, and NIPS seemed like the ideal venue for an auction.

Two months earlier, Hinton and his students had changed the way machines saw the world. They had built what was called a *neural network,* a mathematical system modeled on the web of neurons in the brain, and it could identify common objects—like flowers, dogs, and cars—with an accuracy that had previously seemed impossible. As Hinton and his students showed, a neural network could learn this very human skill by analyzing vast amounts of data. He called this "deep learning," and its potential was enormous. It promised to transform not just computer vision but everything from talking digital assistants to driverless cars to drug discovery.

The idea of a neural network dated back to the 1950s, but the early pioneers had never gotten it working as well as they had hoped. By the new millennium, most researchers had given up on the idea, convinced it was a technological dead end and bewildered by the fifty-year-old conceit that these mathematical systems somehow mimicked the human brain. When submitting research papers to academic journals, those who still explored the technology would often disguise it as something else, replacing the words "neural network" with language less likely to offend their fellow scientists. Hinton remained one of the few who believed it would one day fulfill its promise, delivering machines that could not only recognize objects but identify spoken words, understand natural language, carry on a conversation, and maybe even solve problems humans couldn't solve on their own, providing new and more incisive ways of exploring the mysteries of biology, medicine, geology, and other sciences. It was an eccentric stance even inside his own university, which spent years denying his standing request to hire another professor who could work alongside him in this long and winding struggle to build machines that learned on their own. "One crazy person working on this was enough," he says. But in the spring and summer of 2012, Hinton and his two students made a breakthrough: They showed that a neural network could recognize common objects with an accuracy beyond any other technology. With the nine-page paper they unveiled that fall, they announced to the world that this idea was as powerful as Hinton had long claimed it would be.

Days later, Hinton received an email from a fellow AI researcher named Kai Yu, who worked for Baidu, the Chinese tech giant. On the surface, Hinton and Yu had little in common. Born in postwar Britain to a family of monumental scientists whose influence was matched only by their eccentricity, Hinton had studied at Cambridge, earned a PhD in artificial intelligence from the University of Edinburgh, and spent the next thirty years as a professor of computer science. Born thirty years after Hinton, Yu grew up in Communist China, the son of an automobile engineer, and studied in Nanjing and then Munich before moving to Silicon Valley for a job in a corporate research lab. The two were separated by class, age, culture, language, and geography, but they shared an unusual interest: neural networks. They had originally met in Canada at an academic workshop, part of a grassroots effort to revive this nearly dormant area of research across the scientific community and rebrand the idea as "deep learning." Yu was among those who helped spread the new gospel. Returning to China, he took the idea to Baidu, where his research caught the eye of the company's CEO. When that nine-page paper emerged from the University of Toronto, Yu told the Baidu brain trust they should hire Hinton as quickly as possible. With his email, he introduced Hinton to a Baidu vice president, who offered $12 million for just a few years of work.

At first, Hinton's suitors in Beijing felt they had reached an agreement. But Hinton wasn't so sure. In recent months, he'd cultivated relationships inside several other companies, both small and large, including two of Baidu's big American rivals, and they, too, were calling his office in Toronto, asking what it would take to hire both him and his students. Seeing a much wider opportunity, he asked Baidu if he could solicit other offers before accepting the $12 million, and when Baidu agreed, he flipped the situation upside down. Spurred on by his students and realizing that Baidu and its rivals were much more likely to pay enormous sums of money to acquire a company than they were to shell out the same dollars for a few new hires from the world of academia, he created his tiny start-up. He called it DNNresearch in a nod to the "deep neural networks" they specialized in, and he asked a Toronto lawyer how he could maximize the price of a start-up with three employees, no products, and virtually no history. As the lawyer saw it, he had two options: He could hire a professional negotiator and risk angering the companies he hoped would acquire his tiny venture, or he could set up an auction. Hinton chose an auction. In the end, four names joined the bidding for his new company: Baidu, Google, Microsoft, and a two-year-old start-up most of the world had never heard of. This was DeepMind, a London company founded by a young neuroscientist named Demis Hassabis that would grow to become the most celebrated and influential AI lab of the decade.

The week of the auction, Alan Eustace, Google's head of engineering, flew his own twin-engine plane into the airport near the south shore of Lake Tahoe. He and Jeff Dean, Google's most revered engineer, had dinner with Hinton and his students in the restaurant on the top floor of Harrah's, a steak house decorated with a thousand empty wine bottles. It was Hinton's sixty-fifth birthday. As he stood at a bar table and the others sat on high stools, they discussed Google's ambitions, the auction, and the latest research under way at his lab in Toronto. For the Googlers, the dinner was mostly a way of running the rule over Hinton's two young students, whom they had never met. Baidu, Microsoft, and DeepMind also sent representatives to Lake Tahoe for the conference, and others played their own roles in the auction. Kai Yu, the Baidu researcher who'd kicked off the race for Hinton and his students, held his own meeting with the Toronto researchers before the bidding began. But none of the bidders ever gathered in the same place at the same time. The auction played out over email, with most bids arriving via corporate executives elsewhere in the world, from California to London to Beijing. Hinton hid the identity of each bidder from all the rest.

He ran the auction from his hotel room, number 731 in the Harrah's tower, which looked out over the Nevada pines and onto the snowy peaks of the mountains. Each day, he set a time for the next round of bidding, and at the designated hour, he and his two students would gather in his room to watch the bids arrive on his laptop. The laptop sat on a trash can turned upside down on a table at the end of the room's two queen-sized beds, so that Hinton could type while standing up. The bids arrived via Gmail, the online email service operated by Google, just because this was where he kept an email account. But Microsoft didn't like the arrangement. In the days before the auction, the company complained that Google, its biggest rival, could eavesdrop on its private messages and somehow game the bids. Hinton had discussed the same possibility with his students, though for him this was less a serious concern than an arch comment on the vast and growing power of Google. Technically, Google could read any Gmail message. The terms of service said it wouldn't, but the reality was that if it ever violated those terms, no one was likely to know. In the end, both

Hinton and Microsoft set their concerns aside—"We were fairly confident Google wouldn't read our Gmail," he says—and though no one quite realized it at the time, it was a moment loaded with meaning.

The auction rules were simple: After each bid, the four companies had an hour to raise the buying price by at least a million dollars. This hour-long countdown started at the time stamped on the email holding the latest bid, and at the end of the hour, if no one lodged a new bid, the auction was over. DeepMind bid with company shares, not cash, but it couldn't compete with the giants of the tech world and soon dropped out. That left Baidu, Google, and Microsoft. As the bids continued to climb, first to $15 million and then to $20 million, Microsoft dropped out, too, but then it came back in. Each small moment seemed large and heavy with meaning as Hinton and his students debated which company they'd rather join. Late one afternoon, as they looked out the window at the peaks of the ski mountains, two airplanes flew past from opposite directions, leaving contrails that crossed in the sky like a giant *X*. Amid the atmosphere of excitement in the room, they wondered what this meant, before remembering that Google was headquartered in a place called Mountain View. "Does that mean we should join Google?" Hinton asked. "Or does it mean we shouldn't?"

At about $22 million, Hinton temporarily suspended the auction for a discussion with one of the bidders, and half an hour later, Microsoft dropped out again. That left Baidu and Google, and as the hours passed, the two companies took the price still higher. Kai Yu handled the initial Baidu bids, but when the price reached $24 million, a Baidu executive took over from Beijing. From time to time, Yu would stop by room 731, hoping to glean at least a small sense of where the auction was headed.

Though Yu wasn't even remotely aware of it, these visits were a problem for Hinton. He was sixty-five years old, and he often got sick when he traveled to Lake Tahoe, where the air was cold, thin, and dry. He was worried he might get sick again, and he didn't want Yu, or anyone else, to see him that way. "I didn't want them thinking I was old and decrepit," he says. So he removed the mattress from the pullout couch against the wall, laid it on the floor between the two beds, stretched an ironing board and a few other sturdy objects across the gap between the beds, dampened several towels with water, laid them across the gap, too, and slept each night in the wet air under this makeshift canopy. This, Hinton thought, would keep the illness at bay. The trouble was that as the auction continued, Yu, a small, round-raced man with glasses, kept dropping in for a chat. Now Hinton didn't want Yu to see how determined he was not to get sick. So each time Yu stopped by, Hinton turned to his two students, the only other people in his three-person company, and asked them to hide the mattress and the ironing board and the wet towels. "This is what vice presidents do," he told them.

After one visit, Yu left the room without his backpack, and when Hinton and his students noticed it sitting on a chair, they wondered if they should open it to see if anything inside would tell them how high Baidu was willing to bid. But they didn't, feeling it just wasn't the right thing to do. In any case, they soon realized Baidu was willing to go much higher: $25 million, $30 million, $35 million. Inevitably, the next bid wouldn't arrive until a minute or two before the top of the hour, extending the auction just as it was on the verge of ending.

The price climbed so high, Hinton shortened the bidding window from an hour to thirty minutes. The bids quickly climbed to $40 million, $41 million, $42 million, $43 million. "It feels like we're in a movie," he said. One evening, close to midnight, as the price hit $44 million, he suspended the bidding again. He needed some sleep.

The next day, about thirty minutes before the bidding was set to resume, he sent an email saying the start would be delayed. About an hour later, he sent another. The auction was over. At some point during the night, Hinton had decided to sell his company to Google—without pushing the price any higher. His email to Baidu said that any other messages the company sent would be forwarded to his new employer, though he didn't say who that was.

This, he later admitted, was what he had wanted all along. Even Kai Yu had guessed that Hinton would end up at Google, or at least another American company, because his back would keep him from traveling to China. As it was, Yu was happy just for Baidu to have taken its place among the bidders. By pushing its American rivals to the limit, he believed, the Baidu brain trust had come to realize how important deep learning would be in the years ahead.

Hinton stopped the auction because finding the right home for his research was ultimately more important to him than commanding the maximum price. When he told the bidders at Google he was stopping the auction at $44 million, they thought he was joking—that he couldn't possibly give up the dollars that were still coming. He wasn't joking, and his students saw the situation much as he did. They were academics, not entrepreneurs, more loyal to their idea than to anything else.

But Hinton didn't realize how valuable their idea would prove to be. No one did. Alongside a small group of other scientists—spread across those same four companies, one more American Internet giant, and, eventually, a new upstart—Hinton and his students soon pushed this single idea into the heart of the tech industry. In doing so, they suddenly and dramatically accelerated the progress of artificial intelligence, including talking digital assistants, driverless cars, robotics, automated healthcare, and—though this was never their intention—automated warfare and surveillance. "It changed the way that I looked at technology," Alan Eustace says. "It changed the way many others looked at it, too."

Some researchers, most notably Demis Hassabis, the young neuroscientist behind DeepMind, even believed they were on their way to building a machine that could do *anything* the human brain could do, only better, a possibility that has captured the imagination since the earliest days of the computer age. No one was quite sure when this machine would arrive. But even in the near term, with the rise of machines that were still a very long way from true intelligence, the social implications were far greater than anyone realized. Powerful technologies have always fascinated and frightened humanity, and humanity has gambled on them time and again. This time, the stakes were higher than the scientists behind the idea could know. The rise of deep learning marked a fundamental change in the way digital technology was built. Rather than carefully defining how a machine was supposed to behave, one rule at a time, one line of code at a time, engineers were beginning to build machines that could learn tasks through their own experiences, and these experiences spanned such enormous amounts of digital information, no human could ever wrap their head around it all. The result was a new breed of machine that was not only more powerful than before but also more mysterious and unpredictable.

As Google and other tech giants adopted the technology, no one quite realized it was learning the biases of the researchers who built it. These researchers were mostly white men, and they didn't see the breadth of the problem until a new wave of researchers—both women and people of color—put a finger on it. As the technology spread even further, into healthcare, government surveillance, and the military, so did the ways it might go wrong. Deep learning brought a power even its designers did not completely understand how to control, particularly as it was embraced by tech superpowers driven by an insatiable hunger for revenue and profit.

After Hinton's auction played out in Lake Tahoe and the NIPS conference came to an end, Kai Yu boarded a plane for Beijing. There, he ran into a China-born Microsoft researcher named Li Deng, who had his own history with Hinton and had played his own role in the auction. Yu and Deng knew each other from years of AI conferences and workshops, and they arranged for adjacent seats on the long flight to Asia. Because Hinton kept the bidders anonymous, neither was quite sure which companies were involved in the auction. They wanted to know, and Deng loved to talk. They spent hours standing in the back of the cabin, discussing the rise of deep learning. But they also felt bound by their employers not to reveal their own involvement in the auction. So they danced around the issue, trying to understand what the other knew without giving their own secrets away. Though they didn't say it, both knew that a new competition was on. Their companies would have to answer Google's big move. That was how the tech industry worked. It was the beginning of a global arms race, and this race would quickly escalate in ways that would have seemed absurd a few years before.

Meanwhile, Geoff Hinton took the train back to Toronto. He would eventually make his way to the Google headquarters in Mountain View, California, but even as he joined the company, he retained his professorship at the University of Toronto and held tight to his own aims and beliefs, setting an example for the many other academics who soon followed him into the world's largest tech companies. Years later, when asked to reveal the companies that bid for his start-up, he answered in his own way. "I signed contracts saying I would never reveal who we talked to. I signed one with Microsoft and one with Baidu and one with Google," he said. "Best not to go into that." He did not mention DeepMind. But that was another story. In the wake of the auction in Lake Tahoe, Demis Hassabis, the founder of the London lab, imprinted his own views on the world. In some ways, they echoed Hinton's. In other ways, they looked even further into the future. Soon, Hassabis, too, was caught up in the same global arms race.

This is the story of Hinton and Hassabis and the other scientists who sparked this race, a small but eclectic group of researchers from around the globe who nurtured an idea for decades, often in the face of unfettered skepticism, before it suddenly came of age and they were sucked into the machinery of some of the largest companies on Earth—and a world of turmoil none of them expected.

PART ONE

A NEW KIND OF MACHINE

1

GENESIS

"FRANKENSTEIN MONSTER DESIGNED BY NAVY THAT THINKS."

On July 7, 1958, several men gathered around a machine inside the offices of the United States Weather Bureau in Washington, D.C., about fifteen blocks west of the White House. As wide as a kitchen refrigerator, twice as deep, and nearly as tall, the machine was just one piece of a mainframe computer that fanned across the room like a multipiece furniture set. It was encased in silvery plastic, reflecting the light from above, and the front panel held row after row of small round lightbulbs, red square buttons, and thick plastic switches, some white and some gray. Normally, this $2 million machine ran calculations for the Weather Bureau, the forerunner of the National Weather Service, but on this day, it was on loan to the U.S. Navy and a twenty-nine-year-old Cornell University professor named Frank Rosenblatt.

As a newspaper reporter looked on, Rosenblatt and his Navy cohorts fed two white cards into the machine, one marked with a small square on the left, the other marked on the right. Initially, the machine couldn't tell them apart, but after it read another fifty cards, that changed. Almost every time, it correctly identified where the card was marked—left or right. As Rosenblatt explained it, the machine had learned this skill on its own, thanks to a mathematical system modeled on the human brain. He called it a Perceptron. In the future, he said, this system would learn to recognize printed letters, handwritten words, spoken commands, and even people's faces, before calling out their names. It would translate one language into another. And in theory, he added, it could clone itself on an assembly line, explore distant planets, and cross the line from computation into sentience.

"The Navy revealed the embryo of an electronic computer today that it expects will be able to walk, talk, see, write, reproduce itself, and be conscious of its existence," read the article that appeared the next morning in the *New York Times*. A second article, in the Sunday edition, said that Navy officials hesitated to call this a machine because it was "so much like a human being without life." Rosenblatt grew to resent the way the popular press covered the event, particularly a headline in Oklahoma ("Frankenstein Monster Designed by Navy That Thinks"). In later years, among colleagues and in his published writings, he described the project in more measured terms. He insisted it was not an attempt at artificial intelligence, and he acknowledged its limitations. Still, the idea slipped from his grasp.

The Perceptron was one of the first neural networks, an early incarnation of the technology Geoff Hinton would auction to the highest bidder more than fifty years later. But before it reached that $44 million moment, let alone the extravagant future predicted across the pages of the *New York Times* in the summer of 1958, it descended into academic obscurity. By the early 1970s, after those lavish predictions met the limitations of Rosenblatt's technology, the idea was all but dead.

FRANK Rosenblatt was born on July 11, 1928, in New Rochelle, New York, just north of the Bronx. He attended Bronx Science, the elite public high school that eventually produced eight Nobel laureates, six Pulitzer Prize winners, eight National Medal of Science winners, and three recipients of the Turing Award, the world's top computer science prize. A small, thin man with fleshy jowls and short, dark, wavy hair who wore standard-issue black-rimmed glasses, Rosenblatt was trained in psychology, but his interests were much wider. In 1953, the *New York Times* published a small story describing an early computer he used to crunch data for his PhD thesis. Called EPAC—short for "electronic profile-analyzing computer"—it analyzed the psychological profiles of his patients. As the years passed, he came to believe that machines could provide an even greater understanding of the mind. After finishing his PhD, he joined the Cornell Aeronautical Laboratory in Buffalo, about a hundred and fifty miles from the university's main campus in Ithaca, New York. Donated to Cornell by a company that designed aircraft during the Second World War, this flight research center morphed into a more eclectic lab in the postwar years, operating with little oversight from the administration back in Ithaca. It was here that Rosenblatt designed the Perceptron, backed by funding from the Office of Naval Research.

Rosenblatt saw the project as a window into the inner workings of the brain. If he could re-create the brain as a machine, he believed, he could plumb the mysteries of what he called "natural intelligence." Drawing on ideas initially proposed by two researchers at the University of Chicago a decade earlier, the Perceptron analyzed objects and looked for patterns that could identify these objects (for example, whether a card had a mark on the left-hand or right-hand side). It did this using a series of mathematical calculations that operated (in a very broad sense) like the network of neurons in the brain. As the Perceptron examined and tried to identify each object, it got some right and others wrong. But it could learn from its mistakes, methodically adjusting each of those mathematical calculations until the mistakes were few and far between. Much like a neuron in the brain, each calculation was nearly meaningless on its own—just an input for a larger algorithm. But the larger algorithm—a kind of mathematical recipe—could actually do something useful. Or at least that was the hope. At the Weather Bureau in the summer of 1958, Rosenblatt showed off the beginnings of this idea—a simulation of the Perceptron that ran on the bureau's IBM 704, the leading commercial computer of the day. Then, back at the lab in Buffalo, working alongside a team of engineers, he began building an entirely new machine around the same idea. He called it the Mark I. Unlike other machines of the day, it was designed to see the world around it. "For the first time, a nonbiological system will achieve an organization of its external environment in a meaningful way," he told a reporter later that year, during another trip to meet with his backers in Washington.

His primary collaborator at the Office of Naval Research did not view the Perceptron in the same extravagant terms. But Rosenblatt was unmoved. "My colleague disapproves of all the loose talk one hears nowadays about mechanical brains," he told the reporter, over a cup of coffee. "But that is exactly what it is." A small, silver pitcher of cream sat on the table in front of him, and he picked it up. Though this was the first time he'd laid eyes on the pitcher, Rosenblatt said, he could still recognize it as a pitcher. The Perceptron, he explained, could do much the same. It could draw the conclusions needed to distinguish, say, a dog from a cat. He admitted that the technology was a long way from having practical uses: It lacked depth perception and "the refinements of judgment." But he was confident of its potential. One day, he said, the Perceptron would travel into space and send its observations back to Earth. When the reporter asked if there was anything the Perceptron was *not* capable of, Rosenblatt threw up his hands. "Love. Hope. Despair. Human nature, in short," he said. "If we don't understand the human sex drive, how should we expect a machine to?"

That December, the *New Yorker* hailed Rosenblatt's creation as the first serious rival to the brain. Previously, the magazine had marveled that the IBM 704 could play a game of chess. Now it described the Perceptron as an even more remarkable machine, a computer that could achieve "what amounts to human thought." Though scientists claimed that only biological systems could see, feel, and think, the magazine said, the Perceptron behaved *"as if* it saw, felt, and thought." Rosenblatt had not yet built the machine, but this was dismissed as a minor obstacle. "It is only a question of time (and money) before it comes into existence," the magazine said.

Rosenblatt completed the Mark I in 1960. It spanned six racks of electrical equipment, each the size of a kitchen fridge, and it plugged into what seemed to be a camera. This *was* a camera, though engineers had removed the film loader, swapping in a small square device covered with four hundred black dots. These were photocells that responded to changes in light. Rosenblatt and his engineers would print block letters on squares of cardboard—*A, B, C, D,* etc.—and when they placed these squares on an easel in front of the camera, the photocells could read the black lines of the letters against the white space of the cardboard. As they did, the Mark I learned to recognize the letters, much like the IBM mainframe inside the Weather Bureau learned to recognize the marked cards. This required a little help from the humans in the room: As it worked to identify the letters, a technician would tell the machine whether it was right or wrong. But eventually, the Mark I would learn from its own hits and misses, pinpointing the patterns that identified the slanted line of an *A* or the double curve of a *B*. When demonstrating the machine, Rosenblatt had a way of proving this was learned behavior. He would reach into the racks of electrical equipment and pull out a few wires, breaking the connections between the motors that acted as faux-neurons. When he reconnected the wires, the machine once again struggled to recognize the letters, but then, after examining more cards and relearning the same skill, it worked as it did before.

This electrical contraption worked well enough to attract interest beyond the Navy. Over the next several years, the Stanford Research Institute, or SRI, a lab in Northern California, began exploring the same ideas, and Rosenblatt's own lab won contracts with both the U.S. Postal Service and the Air Force. The Postal Service needed a way of reading addresses on envelopes, and the Air Force hoped to identify targets in aerial photos. But all that was still in the future. Rosenblatt's system was only marginally effective when reading printed letters, a relatively simple task. As the system analyzed cards printed with the letter *A,* each photocell examined a particular spot on the card—say, an area near the lower right-hand corner. If the spot was black more often than white, the Mark I assigned it a high "weight," meaning it would play a more important role in the mathematical calculation that ultimately determined what was an *A* and what wasn't. When reading a new card, the machine could recognize an *A* if most of the highly weighted spots were colored black. That was about it. The technology wasn't nearly nimble enough to read the irregularities of handwritten digits.

Despite the system's obvious deficiencies, Rosenblatt remained optimistic about its future. Others, too, believed the technology would improve in the years to come, learning more complex tasks in more complex ways. But it faced a significant obstacle: Marvin Minsky.

FRANK Rosenblatt and Marvin Minsky had been contemporaries at Bronx Science. In 1945, Minsky's parents moved him to Phillips Andover, the model American prep school, and after the war, he enrolled at Harvard. But he complained that neither could match his experience at Science, where the coursework had been more challenging and the students more ambitious—"people you could discuss your most elaborate ideas with and no one would be condescending," he said. After Rosenblatt died, Minsky pointed to his old schoolmate as the kind of creative thinker who walked the halls of Science. Like Rosenblatt, Minsky was a pioneer in the field of artificial intelligence, but he viewed the field through a different lens.

As an undergraduate at Harvard, using over three thousand vacuum tubes and a few parts from an old B-52 bomber, Minsky built what may have been the first neural network, a machine he called SNARC. Then, as a graduate student in the early '50s, he continued to explore the mathematical concepts that eventually gave rise to the Perceptron. But he came to see artificial intelligence as a larger endeavor. He was among the small group of scientists who crystalized AI as its own field of study during a gathering at Dartmouth College in the summer of 1956. A Dartmouth professor named John McCarthy had urged the wider academic community to explore an area of research he called "automata studies," but that didn't mean much to anyone else. So he recast it as "artificial intelligence," and that summer, he organized a conference alongside several like-minded academics and other researchers. The agenda at the Dartmouth Summer Research Conference on Artificial

Intelligence included "neuron nets," but also "automatic computers," "abstractions," and "self-improvement." Those who attended the conference would lead the movement into the 1960s, most notably McCarthy, who eventually took his research to Stanford University on the West Coast; Herbert Simon and Alan Newell, who built a lab at Carnegie Mellon in Pittsburgh; and Minsky, who settled at the Massachusetts Institute of Technology in New England. They aimed to re-create human intelligence using whatever technology could get them there, and they were sure it wouldn't take very long, some arguing that a machine would beat the world chess champion and discover its own mathematical theorem within a decade. Bald from a young age, with wide ears and an impish grin, Minsky became an AI evangelist, but his evangelism didn't extend to neural networks. A neural network was just one way of building artificial intelligence, and Minsky, like many of his colleagues, began exploring other avenues. By the mid-'60s, as other techniques grabbed his attention, he questioned whether neural networks could handle anything beyond the simple tasks Rosenblatt demonstrated at his lab in upstate New York.

Minsky was part of a larger backlash against Rosenblatt's ideas. As Rosenblatt himself wrote in his 1962 book *Principles of Neurodynamics*, the Perceptron was a controversial concept among academics, and he placed much of the blame on the press. The reporters who wrote about his work in the late 1950s, Rosenblatt said, "fell to the task with all of the exuberance and sense of discretion of a pack of happy bloodhounds." He lamented, in particular, headlines like the one in Oklahoma, saying they were a long way from inspiring confidence in his work as a serious scientific pursuit. Four years after the event in Washington, pulling back on his own early claims, he insisted the Perceptron was not an attempt at artificial intelligence—at least, not as researchers like Minsky saw AI. "The perceptron program is not primarily concerned with the invention of devices for 'artificial intelligence,' but rather with investigating the physical structures and neurodynamic principles which underlie 'natural intelligence,'" he wrote. "Its utility is in enabling us to determine the physical conditions for the emergence of various psychological properties." In other words, he wanted to understand how the human brain worked, rather than send a new brain out into the world. Because the brain was a mystery, he couldn't re-create it. But he believed he could use machines to explore this mystery, and maybe even solve it.

From the beginning, the lines that separated artificial intelligence from computer science, psychology, and neuroscience were blurred as various academic camps sprung up around this new breed of technology, each mapping the landscape in its own way. Some psychologists, neuroscientists, and even computer scientists came to see machines in the way Rosenblatt did: as a reflection of the brain. Others looked on this grandiose idea with scorn, arguing that computers operated nothing like the brain, and that if they were ever going to mimic intelligence, they would have to do it in their own way. No one, however, was anywhere close to building what could rightly be called "artificial intelligence." Though the field's founding fathers thought the path to re-creating the brain would be a short one, it turned out to be very long. Their original sin was that they called their field "artificial intelligence." This gave decades of onlookers the impression that scientists were on the verge of re-creating the powers of the brain when, in reality, they were not.

In 1966, a few dozen researchers traveled to Puerto Rico, gathering at the Hilton hotel in San Juan. They met to discuss the latest advances in what was then called "pattern recognition"—technology that could identify patterns in images and other data. Whereas Rosenblatt viewed the Perceptron as a model of the brain, others saw it as a means of pattern recognition. In later years, some commentators imagined Rosenblatt and Minsky locking horns at academic conferences like the one in San Juan, openly debating the future of the Perceptron, but their rivalry was implicit. Rosenblatt didn't even travel to Puerto Rico. Inside the Hilton, the tension emerged when a young scientist named John Munson addressed the conference. Munson worked at SRI, the Northern California lab that embraced Rosenblatt's ideas after the arrival of the Mark I. There, alongside a larger team of researchers, he was trying to build a neural network that could read handwritten characters, not just printed letters, and with his presentation at the conference, he aimed to show the progress of this research. But when Munson finished the lecture and took questions from the floor, Minsky made himself heard. "How can an intelligent young man like you," he asked, "waste your time with something like this?"

Sitting in the audience, Ron Swonger, an engineer from the Cornell Aeronautical Lab, the birthplace of the Mark I, was appalled. He bristled at Minsky's language, and he questioned whether the attack had anything to do with the presentation delivered from the front of the room. Minsky wasn't concerned with recognizing handwritten characters. He was attacking the very idea of the Perceptron. "This is an idea with no future," he said. Across the room, Richard Duda, who was part of the team trying to build the system for handwritten characters, was stung by the laughter from the audience as Minsky made light of claims that the Perceptron mirrored the network of neurons in the brain. The performance was typical of Minsky, who enjoyed stirring public controversy. He once greeted a roomful of physicists by saying that the field of artificial intelligence had made more progress in just a few years than physics had made in centuries. But Duda also felt the MIT professor had practical reasons for attacking the work at places like SRI and Cornell: MIT was competing with these labs for the same government research dollars. Later in the conference, when another researcher presented a new system designed to create computer graphics, Minsky praised its ingenuity—and took another swipe at Rosenblatt's ideas. "Can a Perceptron do that?" he said.

In the wake of the conference Minsky and an MIT colleague named Seymour Papert published a book on neural networks, which they titled *Perceptrons*. Many felt it shut the door on Rosenblatt's ideas for the next fifteen years. Minsky and Papert described the Perceptron in elegant detail, exceeding, in many respects, the way Rosenblatt described it himself. They understood what it could do, but they also understood its flaws. The Perceptron, they showed, couldn't deal with what mathematicians called "exclusive-or," an esoteric concept that carried much larger implications. When presented with two spots on a cardboard square, the Perceptron could tell you if both were colored black. And it could tell you if both were white. But it couldn't answer the straightforward question: "Are they *two different colors*?" This showed that in some cases, the Perceptron couldn't recognize simple patterns, let alone the enormously complex patterns that characterized aerial photos or spoken words. Some researchers, Rosenblatt among them, were already exploring a new kind of Perceptron that aimed to fix this flaw. Still, in the wake of Minsky's book, the government dollars moved into other technologies, and Rosenblatt's ideas faded from view. Following Minsky's lead, most researchers embraced what was called "symbolic AI."

Frank Rosenblatt aimed to build a system that learned behavior on its own in the same way the brain did. In later years, scientists called this "connectionism," because, like the brain, it relied on a vast array of interconnected calculations. But Rosenblatt's system was much simpler than the brain, and it learned only in small ways.

Like other leading researchers in the field, Minsky believed that computer scientists would struggle to re-create intelligence unless they were willing to abandon the strictures of that idea and build systems in a very different and more straightforward way. Whereas neural networks learned tasks on their own by analyzing data, symbolic AI did not. It behaved according to very particular instructions laid down by human engineers—discrete rules that defined everything a machine was supposed to do in each and every situation it might encounter. They called it symbolic AI because these instructions showed machines how to perform specific operations on specific collections of *symbols,* such as digits and letters. For the next decade, this is what dominated AI research. The movement reached the height of its ambition in the mid-1980s with a project called Cyc, an effort to re-create common sense one logical rule at a time. A small team of computer scientists, based in Austin, Texas, spent their days recording basic truths like "You can't be in two places at the same time" and "When drinking a cup of coffee, you hold the open end up." They knew this would take decades, perhaps centuries. But like so many others, they believed it was the only way.

Rosenblatt tried pushing the Perceptron beyond images. Back at Cornell, he and other researchers built a system for recognizing spoken words. They called it "Tobermory," after a British short story about a talking cat, and it never really worked. By the late '60s, Rosenblatt shifted to a very different area of research, running brain experiments on rats. After one group of rats learned to navigate a maze, he would inject their brain matter into a second group. Then he would drop the second group into the maze to see if their minds had absorbed what the first group had learned. The results were inconclusive.

In the summer of 1971, on his forty-third birthday, Rosenblatt died in a boating accident on the Chesapeake Bay. The newspapers didn't say what happened out on the water. But according to a colleague, he took two students out into the bay on a sailboat. The students had never sailed before, and when the boom swung into Rosenblatt, knocking him into the water, they didn't know how to turn the boat around. As he drowned in the bay, the boat kept going.

2

PROMISE

"OLD IDEAS ARE NEW."

One afternoon in the mid-1980s, a group of nearly twenty academics gathered at an old French manor–style estate just outside Boston that served as a nearby retreat for professors and students from the Massachusetts Institute of Technology, the university where Marvin Minsky still reigned over the international community of AI researchers. They sat at a big wooden table in the center of the room, and as Geoff Hinton walked around the table, he handed each of them a long, rhetorical, math-laden academic paper describing something he called the Boltzmann Machine. Named for the Austrian physicist and philosopher, this was a new kind of neural network, designed to overcome the flaws Minsky pinpointed in the Perceptron fifteen years earlier. Minsky removed the staple from his copy of the paper and spread the pages across the table in front of him, one beside the next, and he looked down on this long line of pages as Hinton returned to the front of the room and gave a brief lecture explaining his new mathematical creation. Minsky didn't speak. He just looked. Then, when the lecture ended, he stood up and walked out of the room, leaving the pages behind, still lined up in neat rows on the table.

Although neural networks had fallen from favor in the wake of Minsky's book on the Perceptron, Hinton, then a computer science professor at Carnegie Mellon University in Pittsburgh, had kept the faith, building the Boltzmann Machine in collaboration with a researcher named Terry Sejnowski, a neuroscientist at Johns Hopkins in Baltimore. They were part of what one contemporary later called "the neural network underground." The rest of the AI movement was focused on symbolic methods, including the Cyc project under way in Texas. Hinton and Sejnowski, in contrast, believed the future still lay in systems that could learn behavior on their own. The Boston meeting gave them the opportunity to share their new research with the wider academic community.

For Hinton, Minsky's response was typical of the man. He had first met the MIT professor five years earlier, and he had come to see him as immensely curious and creative, but also strangely childlike and vaguely irresponsible. Hinton often told the story of the time Minsky taught him how to make "perfect black"—a color with no color at all. You couldn't make perfect black with pigments, Minsky explained, because they always reflected the light. But you could make it with several tiers of razor blades arranged into V shapes, so that the light would stream into the V, bounce endlessly among the blades, and never escape. Minsky didn't actually demonstrate the trick, and Hinton never tried it. It was classic Minsky—fascinating and thought-provoking, but seemingly casual and untested. And it suggested that he didn't always say what he believed. Certainly, when it came to neural networks, Minsky might have attacked them as woefully deficient—and written a book that many held up as proof they were a dead end—but his true stance wasn't necessarily as clear-cut. Hinton saw him as a "lapsed neural-net-er," someone who had once embraced the notion of machines that behaved like the network of neurons in the brain, grew disillusioned when the idea didn't live up to his expectations, but still held at least some hope that they would fulfill their promise. After Minsky left the lecture in Boston, Hinton gathered the unstapled pages from across the desk and mailed them to Minsky's office, including a short note. It read: "You may have left these behind by accident."

GEOFFREY Everest Hinton was born in Wimbledon, England, just after the Second World War. He was the great-great-grandson of both George Boole, the nineteenth-century British mathematician and philosopher whose "Boolean logic" would provide the mathematical foundation for every modern computer, and James Hinton, the nineteenth-century surgeon who wrote a history of the United States. His great-grandfather was Charles Howard Hinton, the mathematician and fantasy writer whose notion of the fourth dimension, including what he called the "tesseract," ran through popular science fiction for the next one hundred and thirty years, reaching a pop cultural peak with the Marvel superhero movies in the 2010s. His great-uncle Sebastian Hinton invented the jungle gym. And his cousin, the nuclear physicist Joan Hinton, was one of the few women to work on the Manhattan Project. In London and later in Bristol, he grew up alongside three siblings, a mongoose, a dozen Chinese turtles, and two viper snakes that lived in a pit at the back of the garage. His father was the entomologist Howard Everest Hinton, a Fellow of the Royal Society, whose interest in wildlife extended beyond insects. Like his father, he was named for another relative, Sir George Everest, the surveyor general of India whose name was also given to the world's highest peak. It was expected that he would one day follow his father into academia. It was less clear what he would study.

He wanted to study the brain. His interest, he liked to say, was sparked while he was still a teenager, when a friend told him the brain worked like a hologram, storing bits and pieces of memories across a network of neurons, much as a hologram stored bits and pieces of its 3-D image across a strip of film. It was a simple analogy, but the idea drew him in. As an undergraduate at King's College, Cambridge, what he wanted was a better understanding of the brain. The trouble, he soon realized, was that no one understood much more than he did. Scientists understood certain parts of the brain, but they knew very little about how all those parts fit together and ultimately delivered the ability to see, hear, remember, learn, and think. Hinton tried physiology and chemistry, physics and psychology. None offered the answers he was looking for. He started a degree in physics but dropped out, convinced his math skills weren't strong enough, before switching to philosophy. Then he dropped philosophy for experimental psychology. In the end, despite pressure to continue his studies—or perhaps because of that pressure, brought down by his father—Hinton left academia entirely. When he was a boy, he had seen his father as both an uncompromising intellectual and a man of enormous strength—a Fellow of the Royal Society who could do a pull-up with one arm. "Work really hard and maybe when you're twice as old as me, you'll be half as good," his father had often told him, without irony. After graduating from Cambridge, with thoughts of his father hanging over him, Hinton moved to London and became a carpenter. "It wasn't fancy carpentry," he says. "It was carpentry to make a living."

That year, he read *The Organization of Behavior*, a book written by a Canadian psychologist named Donald Hebb that aimed to explain the basic biological process that allowed the brain to learn. Learning, Hebb believed, was the result of tiny electrical signals that fired along a series of neurons, causing a physical change that wired these neurons together in a new way. As his disciples said: "Neurons that fire together, wire together." This theory—known as Hebb's Law—helped inspire the artificial neural networks built by scientists like Frank Rosenblatt in the 1950s. It also inspired Geoff Hinton. Every Saturday, he would carry a notebook to his local public library in Islington, North London, and spend the morning filling its pages with his own theories of how the brain must work, building on the ideas laid down by Hebb. These Saturday-morning scribbles were meant for no one but himself, but they eventually led him back to academia. They happened to coincide with the first big wave of investment in artificial intelligence from the British government and the rise of a graduate program at the University of Edinburgh.

In these years, the cold reality was that neuroscientists and psychologists had very little understanding of how the brain worked, and computer scientists were nowhere close to mimicking the behavior of the brain. But much like Frank Rosenblatt before him, Hinton came to believe that each side—the biological and the artificial—could help the other move forward. He saw artificial intelligence as a way of testing his theories of how the brain worked and, ultimately, understanding its mysteries. If he could understand those mysteries, he could, in turn, build more powerful forms of AI. After a year of carpentry in London, he took a short-term job on a psychology project at the University of Bristol, where his father taught, and he used this as a springboard to the AI program in Edinburgh. Years later, a colleague introduced him at an academic conference as someone who had failed at physics, dropped out of psychology, and then joined a field with no standards at all: artificial intelligence. It was a story Hinton enjoyed repeating, with a caveat. "I didn't fail at physics and drop out of psychology," he would say. "I failed at psychology and dropped out of physics—which is far more reputable."

At Edinburgh, he won a spot in a lab overseen by a researcher named Christopher Longuet-Higgins. Longuet-Higgins had been a theoretical chemist at Cambridge and a rising star in the field, but in the late 1960s, he was drawn to the idea of artificial intelligence. So he left Cambridge for Edinburgh and embraced a breed of AI not unlike the methods that underpinned the Perceptron. His connectionist approach dovetailed with the theories Hinton scribbled in his notebook at the Islington library. But this intellectual harmony was fleeting. In between Hinton accepting a spot in the lab and the day he actually arrived, Longuet-Higgins had another change of heart. After reading Minsky and Papert's book on the Perceptron and a thesis on natural language systems from one of Minsky's students at MIT, he abandoned brainlike architectures and switched to symbolic AI—an echo of the shift across the field as a whole. This meant Hinton spent his graduate school years working in an area that was dismissed not only by his colleagues but by his own advisor. "We met once a week," Hinton says. "Sometimes it ended in a shouting match, sometimes not."

Hinton had very little experience with computer science, and he wasn't all that interested in mathematics, including the linear algebra that drove neural networks. He sometimes practiced what he called "faith-based differentiation." He would dream up an idea, including the underlying differential equations, and just assume the math was right, letting someone else struggle through the calculations needed to ensure that it actually was, or deigning to solve the equations himself when he absolutely had to. But he had a clear belief in how the brain worked—and how machines could mimic the brain. When he told anyone in the field he was working on neural networks, they inevitably mentioned Minsky and Papert. "Neural networks have been disproved," they would say. "You should work on something else." But as Minsky and Papert's book pushed most researchers away from connectionism, it drew Hinton closer. He read it during his first year in Edinburgh. The Perceptron described by Minsky and Papert, he felt, was almost a caricature of Rosenblatt's work. They never quite acknowledged that Rosenblatt saw the same flaws in the technology they saw. What Rosenblatt lacked was their knack for describing these limitations, and perhaps because of that, he didn't know how to address them. He wasn't someone who was going to be slowed by an inability to prove his own theories. Hinton felt that by pinpointing the limitations of a neural network with a sophistication that was beyond Rosenblatt, Minsky and Papert ultimately made it easier to move beyond these problems.

But that would take another decade.

THE year Hinton entered the University of Edinburgh, 1971, the British government commissioned a study on the progress of artificial intelligence. It proved to be damning. "Most workers in AI research and in related fields confess to a pronounced feeling of disappointment in what has been achieved in the past twenty-five years," the report said. "In no part of the field have the discoveries made so far produced the major impact that was then promised." So the government cut funding across the field, ushering in what researchers would later call an "AI winter." This was when the hype that built up behind high-minded notions of artificial intelligence ran into the field's modest technological gains, leading concerned government officials to pull back on additional investment, which slowed progress even further. The analogy was a nuclear winter, when soot covers the skies after a nuclear war, blocking the rays of the sun for years on end. By the time Hinton was finishing his thesis, his research was on the fringes of a shrinking field. Then his father died. "The old bastard died before I was successful," Hinton says. "Not only that, he got a cancer with a high genetic linkage. The last thing he did was increase my chances of dying."

After finishing his thesis, and as the AI winter grew colder, Hinton struggled to find a job. Only one university even offered him an interview. He had no choice but to look abroad, including the United States. AI research was on the wane in the U.S., too, as government agencies reached the same conclusions as those in the UK, pulling back on funding at the big universities. But at the southern end of California, much to his own surprise, he found a small group of people who believed in the same ideas he did.

They were called the PDP group. Short for "parallel distributed processing," PDP was another way of saying "perceptrons" or "neural networks" or "connectionism." It was also a pun, of sorts. In those years—the late 1970s—PDP was a computer chip used in some of the industry's most powerful machines. But the academics in the PDP group weren't computer scientists. They didn't even consider themselves AI researchers. Rather, the group spanned several academics in the psychology department at the University of California–San Diego, and at least one neuroscientist—Francis Crick, from the Salk Institute, the biological research center just down the street. Crick had won a Nobel Prize for discovering the structure of the DNA molecule before turning his attention to the brain. In the fall of 1979, he published a call to arms in the pages of *Scientific American* urging the larger scientific community to at least make an attempt at understanding how the brain worked. Hinton, who was now embarking on a postdoc at the university, experienced a kind of academic culture shock. In Britain, academia was an intellectual monoculture. In the United States, the landscape was large enough to accommodate pockets of dissent. "There could be different views," Hinton says, "and they could survive." Now, if he told other researchers he was working on neural networks, they listened.

There was a straight line from Frank Rosenblatt to the research under way in Southern California. In the '60s, Rosenblatt and other scientists hoped to build a new kind of neural network, a system that spanned multiple "layers" of neurons. In the early '80s, this was also the hope in San Diego. The Perceptron was a single-layer network, meaning there was only one layer of neurons between what the network took in (the image of a capital letter printed on a cardboard square) and what it spit out (the *A* it found in the image). But Rosenblatt believed that if researchers could build a multilayered network, each layer feeding information into the next, this system could learn the complex patterns his Perceptron could not—in other words, that a more brainlike system would emerge. When the Perceptron analyzed cards printed with the letter *A*, each neuron examined a spot on the card and learned whether that particular spot was typically part of the three black lines that defined an *A*. But with a multilayered network, this was just a starting point. Give a photo of, say, a dog to this more complex system, and a far more sophisticated analysis would follow. The first layer of neurons would examine each pixel: Is it black or white, brown or yellow? Then the first layer would feed what it learned into the second layer, where another set of neurons would look for patterns in these pixels, like a small line or a tiny arc. A third layer would look for patterns in the patterns. It might piece several lines together and find an ear or a tooth, or combine those tiny arcs and find an eye or a nostril. In the end, this multilayer network could piece together a dog. That, at least, was the idea. No one had actually gotten it to work. In San Diego, they were trying.

One of the leading figures in the PDP group was a San Diego professor named David Rumelhart, who held degrees in both psychology and mathematics. When asked about Rumelhart, Hinton liked to recall the time they were stuck listening to a lecture that held absolutely no interest for either of them. When the lecture ended, as Hinton complained that he had just lost an hour of his life, Rumelhart said he didn't really mind. If he just ignored the lecture, Rumelhart said, he had sixty uninterrupted minutes to think about his own research. For Hinton, this epitomized his longtime collaborator.

Rumelhart had set himself a very particular, but central, challenge. One of the great problems with building a multilayered neural network was that it was very difficult to determine the relative importance ("the weight") of each neuron to the calculation as a whole. With a single-layer network, like the Perceptron, this was at least doable: The system could automatically set its own weights across its single layer of neurons. But with a multilayered network, such an approach simply didn't work. The relationships between the neurons were too expansive and too complex. Changing the weight of one neuron meant changing all the others that depended on its behavior. A more powerful mathematical method was needed where each weight was set in conjunction with all the others. The answer, Rumelhart suggested, was a process called "backpropagation." This was essentially an algorithm, based on differential calculus, that sent a kind of mathematical feedback cascading down the hierarchy of neurons as they analyzed more data and gained a better understanding of what each weight should be.

When Hinton arrived in San Diego, fresh from his PhD, and

they discussed the idea, he told Rumelhart this mathematical trick would never work. After all, he said, Frank Rosenblatt, the man who designed the Perceptron, *had proven it would never work.* If you built a neural network and you set all the weights to zero, the system could learn to adjust them on its own, sending changes cascading down the many layers. But in the end, each weight would wind up at the same place as all the rest. However much you tried to get the system to adopt relative weighting, its natural tendency was to even things out. As Frank Rosenblatt had shown, this was just how the math behaved. In the vernacular of mathematics, the system couldn't "break symmetry." One neuron could never be more important than any other, and that was a problem. It meant that this neural network wasn't any better than the Perceptron.

Rumelhart listened to Hinton's objection. Then he made a suggestion. "What if you didn't set the weights to zero?" he asked. "What if the numbers were random?" If all the weights held different values at the beginning, he suggested, the math would behave differently. It wouldn't even out all the weights. It would find the weights that allowed the system to actually recognize complex patterns, such as a photo of a dog.

Hinton liked to say that "old ideas are new"—that scientists should never give up on an idea unless someone had proven it wouldn't work. Twenty years earlier, Rosenblatt had proven that backpropagation wouldn't work, so Hinton gave up on it. Then Rumelhart made this small suggestion. Over the next several weeks, the two men got to work building a system that began with random weights, and it could break symmetry. It could assign a different weight to each neuron. And in setting these weights, the system could actually recognize patterns in images. These were simple images. The system couldn't recognize a dog or a cat or a car, but thanks to backpropagation, it could now handle that thing called "exclusive-or," moving beyond the flaw that Marvin Minsky pinpointed in neural networks more than a decade earlier. It could examine two spots on a piece of cardboard and answer the elusive question: "Are they two different colors?" Their system didn't do much more than that, and once again, they set the idea aside. But they'd found a way around Rosenblatt's proof.

In the years that followed, Hinton started a separate partnership with Terry Sejnowski, who was then a postdoc in the biology department at Princeton. They met via a second (unnamed) group of connectionists that convened once a year, at various places across the country, to discuss many of the same ideas percolating in San Diego. Backpropagation was one of them. The Boltzmann Machine was another. Years later, when asked to explain the Boltzmann Machine for the benefit of an ordinary person who knew little about math or science, Hinton declined to do so. This, he said, would be like Richard Feynman, the Nobel Prize–winning physicist, explaining his work in quantum electrodynamics. When anyone asked Feynman to explain the work that won him the Nobel Prize in terms the layperson could understand, he, too, would decline. "If I could explain it to the average person," he would say, "it wouldn't have been worth the Nobel Prize." The Boltzmann Machine was certainly hard to explain, in part because it was a mathematical system based on a one-hundred-year-old theory from the Austrian physicist Ludwig Boltzmann that involved a phenomenon that seemed completely unrelated to artificial intelligence (the equilibrium of particles in a heated gas). But the aim was simple: It was a way of building a better neural network.

Like the Perceptron, the Boltzmann Machine would learn by analyzing data, including sounds and images. But it added a new twist. It would also learn by creating its own sounds and images and then comparing what it created to what it analyzed. This was a bit like the way humans thought, in that humans could envision images and sounds and words. They dreamt, both at night and during the day, before using these thoughts and visions as they navigated the real world. With the Boltzmann Machine, Hinton and Sejnowski hoped to re-create this very human phenomenon with digital technology. "It was the most exciting time of my life," Sejnowski says. "We were convinced we had figured out how brains work." But, like backpropagation, the Boltzmann Machine was still ongoing research that didn't quite do anything useful. For years, it, too, lingered on the fringes of academia.

Hinton's religious commitment to a wide range of unpopular ideas may have left him outside the mainstream, but it did lead to a new job. A Carnegie Mellon professor named Scott Fahlman had joined the same yearly connectionist meeting as Hinton and Sejnowski and had come to believe that hiring Hinton was a way for the university to hedge its bets on artificial intelligence. Like MIT and Stanford and most other labs in the world, Carnegie Mellon was focused on symbolic AI. Fahlman saw neural networks as a "crazy idea," but he also acknowledged that the other ideas under development at the university were perhaps just as crazy. In 1981, with Fahlman as his sponsor, Hinton visited the university for a job interview, giving two lectures: one in the psychology department and one in computer science. His lectures were a fire hose of information that gave little pause for the uninitiated as he waved his arms alongside each sentence, spreading his hands far apart before bringing them back together as he made his point. These talks weren't heavy on math or computer science, just because he wasn't all that interested in math or computer science. They were heavy on ideas, and for those who were willing and able to keep up, they were strangely exhilarating. That day, his lecture won the attention of Alan Newell, one of the founding fathers of the AI movement, a leading figure in the decades-long push toward symbolic methods, and the head of the Carnegie Mellon computer science department. The following afternoon, Newell offered him a job in the department, though Hinton stopped him before accepting.

"There is something you should know," Hinton said.

"What's that?" Newell asked.

"I don't actually know any computer science."

"That's okay. We have people here who do."

"In that case, I accept."

"What about salary?" Newell asked.

"Oh, no. I don't care about that," Hinton said. "I am not doing it for the money."

Later, Hinton discovered he was paid about a third less than his colleagues ($26,000 versus $35,000), but he'd found a home for his unorthodox research. He continued work on the Boltzmann Machine, often driving to Baltimore on weekends so he could collaborate with Sejnowski in the lab at Johns Hopkins, and somewhere along the way, he also started tinkering with backpropagation, reckoning it would throw up useful comparisons. He thought he needed something he could compare with the Boltzmann Machine, and backpropagation was as good as anything else. An old idea was new. At Carnegie Mellon, he had more than just the opportunity to explore these two projects. He had better, faster computer hardware. This drove the research forward, allowing these mathematical systems to learn more from more data. The breakthrough came in 1985, a year after the lecture he gave Minsky in Boston. But the breakthrough wasn't the Boltzmann Machine. It was backpropagation.

In San Diego, he and Rumelhart had shown that a multilayered neural network could adjust its own weights. Then, at Carnegie Mellon, Hinton showed that this neural network could actually do something that would impress more than just mathematicians. When he fed it pieces of a family tree, it could learn to identify the various relationships among family members, a small skill that showed it was capable of much more. If he told this neural network that John's mother was Victoria and that Victoria's husband was Bill, it could learn that Bill was John's father. Unbeknownst to Hinton, others in completely separate fields had designed mathematical techniques similar to backpropagation in the past. But unlike those before him, he showed that this mathematical idea had a future, and not just with images but with words. It also held more potential than other AI technologies, because it could learn on its own.

The following year, Hinton married a British academic named Rosalind Zalin, a molecular biologist, whom he had met while doing a postdoc at the University of Sussex in Britain. She believed in homeopathic medicine, which would become a source of tension between the two of them. "For a molecular biologist, believing in homeopathy is not honorable. So, life was difficult," Hinton says. "We had to agree not to talk about it." She was also a confirmed socialist who did not take to Pittsburgh or to the politics of Ronald Reagan's America. But for Hinton this was a fruitful period for his own research. The morning of his wedding, he disappeared for half an hour to mail a package to the editors of *Nature,* one of the world's leading science journals. The package contained a research paper describing backpropagation, written with Rumelhart and a Northeastern University professor named Ronald Williams. It was published later that year.

This was the kind of academic moment that goes unnoticed across the larger world, but in the wake of the paper, neural networks entered a new age of optimism and, indeed, progress, riding a larger wave of AI funding as the field emerged from its first long winter. "Backprop," as researchers called it, was not just an idea.

One of the first practical applications came in 1987. Researchers at the Carnegie Mellon artificial intelligence lab were trying to build a truck that could drive by itself. They began with a royal-blue Chevy shaped like an ambulance. Then they installed a suitcase-sized video camera on the roof and filled the rear chassis with what was then called a "supercomputer"—a machine that juggled data a hundred times faster than the typical business computers of the day. The idea was that this machine, spanning several racks of electrical boards, wires, and silicon chips, would read the images streaming from the rooftop camera and decide how the truck should navigate the road ahead. But that would take some doing. A few graduate students were coding all the driving behavior by hand, one line of software at a time, writing detailed instructions for each situation the truck would encounter on the road. It was a Sisyphean task. By that fall, several years into the project, the car could drive itself no faster than a few inches a second.

Then, in 1987, a first-year PhD student named Dean Pomerleau pushed all that code aside and rebuilt the software from scratch using the ideas proposed by Rumelhart and Hinton.

He called his system ALVINN. The two *N*'s stood for "neural network." When he was done, the truck operated in a new way: It could learn to drive by watching how humans navigated the road. As Pomerleau and his colleagues drove the truck across Pittsburgh's Schenley Park, winding along the asphalt bike paths, it used the images streaming from its rooftop camera as a way of tracking what the drivers were doing. Just as Frank Rosenblatt's Perceptron could learn to recognize letters by analyzing what was printed on squares of cardboard, the truck could learn to steer by analyzing how humans handled each turn in the road. Soon it was navigating Schenley Park on its own. At first, this souped-up blue Chevy, carrying several hundred pounds of computing hardware and electrical equipment, drove at speeds no higher than nine or ten miles an hour. But as it continued to learn with Pomerleau and other researchers behind the wheel, analyzing more images on more roads at higher speeds, it continued to improve. At a time when Middle American families put signs in their car windows that said "Baby on Board" or "Grandma on Board," Pomerleau and his fellow researchers outfitted ALVINN with one that read "Nobody on Board." It was true, at least in spirit. Early one Sunday morning in 1991, ALVINN drove itself from Pittsburgh to Erie, Pennsylvania, at nearly sixty miles an hour. Two decades after Minsky and Papert published their book on the Perceptron, it did the kind of thing they said a neural network couldn't do.

Hinton was not there to see it. In 1987, the year Pomerleau arrived at Carnegie Mellon, he and his wife left the United States for Canada. The reason, he liked to say, was Ronald Reagan. In the United States, the majority of the funding for artificial intelligence research flowed from military and intelligence organizations, most notably the Defense Advanced Research Projects Agency, or DARPA, an arm of the U.S. Defense Department dedicated to emerging technologies. Created in 1958 in response to the Sputnik satellite launched by the Soviet Union, DARPA had funded AI research since the field's earliest days. This was the primary source of the grant money that Minsky pulled away from Rosenblatt and other connectionists in the wake of his book on the Perceptron, and it funded Pomerleau's work on ALVINN. But amid the political climate in the U.S. at the time, including the controversy swirling around the Iran-Contra affair, when Reagan administration officials secretly sold arms to Iran as a way of financing operations against the socialist government in Nicaragua, Hinton grew to resent his dependence on DARPA money, and his wife, Ros, pushed for a move to Canada, saying she could no longer live in the U.S. At the height of the revival in neural network research, Hinton left Carnegie Mellon for a professorship at the University of Toronto.

A few years after this move, as he struggled to find new funding for his research, he wondered if he had made the right decision.

"I should have gone to Berkeley," he told his wife.

"Berkeley?" his wife said. "I would have gone to Berkeley."

"But you said you wouldn't live in the U.S."

"That's not the U.S. It's California."

But the decision had been made. He was in Toronto. It was a move that would shift the future of artificial intelligence—not to mention the geopolitical landscape.

3

REJECTION

"I WAS DEFINITELY THINKING I WAS RIGHT THE WHOLE TIME."

Yann LeCun sat at a desktop computer, wearing a dark blue sweater over a white button-down. The year was 1989, when desktop computers still plugged into monitors the size of microwave ovens, complete with spinning knobs for adjusting color and brightness. A second cable ran from the back of this machine to what looked like a table lamp hung upside down. But it wasn't a lamp. It was a camera. With a knowing grin, the left-handed LeCun grabbed a slip of paper bearing a handwritten phone number—201-949-4038—and pushed it under the camera. As he did, its image appeared on the monitor. When he touched the keyboard, there was a flash at the top of the screen—a hint of rapid-fire calculation—and after a few seconds, the machine read what was written on the paper, displaying the same number in digital form: "201 949 4038."

This was LeNet, a system created by LeCun and, eventually,

named for him. That phone number—201-949-4038—rang his office at the Bell Labs research center in Holmdel, New Jersey, a neo-futuristic mirrored box of a building designed by the Finnish-American architect Eero Saarinen where dozens of researchers explored new ideas under the aegis of telecom giant AT&T. Bell Labs was perhaps the world's most storied research operation, responsible for the transistor, the laser, the Unix computer operating system, and the C programming language. Now LeCun, a baby-faced twenty-nine-year-old computer scientist and electrical engineer from Paris, was developing a new kind of image-recognition system, building on the ideas developed by Geoff Hinton and David Rumelhart a few years earlier. LeNet learned to recognize handwritten numbers by analyzing what was scribbled on the envelopes of dead letters from the U.S. Postal Service. As LeCun fed envelopes into his neural network, it analyzed thousands of examples of each digit—from 0 to 9—and after about two weeks of training, it could recognize each one on its own.

Sitting at his desktop computer inside the Bell Labs complex in Holmdel, LeCun repeated the trick with several more numbers. The last looked like a grade-school art project: The 4 was double-wide, the 6 a series of circles, the 2 a stack of straight lines. But the machine read them all—and read them correctly. Though it needed weeks to learn a task as simple as identifying a phone number or a zip code, LeCun believed this technology would continue to improve as increasingly powerful computer hardware sped up its training, so it could learn from more data in less time. He saw this as a path to machines that could recognize almost anything captured by a camera, including dogs, cats, cars—even faces. Like Frank Rosenblatt nearly forty years earlier, he also believed that as such research continued, machines would learn to listen and talk and maybe even reason like a human. But this went unspoken. "We were thinking it," he says, "but not really saying it." After so many years of researchers claiming artificial intelligence was nigh when it wasn't, the norms of the research community had changed. If you claimed a path to intelligence, you weren't taken seriously. "You don't make claims like this unless you have evidence to justify them," LeCun says. "You build the system, it works, you say 'Look, here is this performance on this dataset,' and even then, nobody believes you. Even when you actually have the evidence and you show that it works, nobody believes you for a bit."

IN October 1975, at the Abbaye de Royaumont, a medieval abbey just north of Paris, the American linguist Noam Chomsky and the Swiss psychologist Jean Piaget debated the nature of learning. Five years later, a book of essays deconstructed this sweeping debate, and Yann LeCun read them as a young engineering student. Eighty-nine pages in, as an aside, the book mentioned the Perceptron, calling it a device "capable of formulating simple hypotheses from regular exposure to raw data," and LeCun was hooked, instantly enamored with the idea of a machine that could learn. Learning, he believed, was inextricable from intelligence. "Any animal with a brain can learn," he often said.

At a time when few researchers paid much attention to neural networks and those that did saw it not as artificial intelligence but as just another form of pattern recognition, LeCun homed in on the idea during his years as an undergraduate at the École Supérieure d'Ingénieurs en Électrotechnique et Électronique. Most of the papers he studied were written in English by Japanese researchers, because Japan was one of the few places where this research still went on. Then he discovered the movement in North America. In 1985, LeCun attended a conference in Paris dedicated to exploring new and unusual approaches to computer science. Hinton was there, too, giving a lecture on the Boltzmann Machine. At the end of his talk, LeCun followed him out the room, convinced he was one of the few people on Earth who held the same beliefs. LeCun couldn't quite reach him in the melee, but then Hinton turned to another man and asked: "Do you know someone called Yann LeCun?" As it turned out, Hinton had heard about the young engineering student from Terry Sejnowski, the other man behind the Boltzmann Machine, who had met LeCun at a workshop a few weeks before. The name had slipped Hinton's mind, but after reading no more than the title of LeCun's research paper in the conference program, he knew this had to be the one.

The next day, the two had lunch at a local North African restaurant. Though Hinton knew almost no French and LeCun knew little English, they had little trouble communicating as they ate their couscous and discussed the vagaries of connectionism. LeCun felt as if Hinton was completing his sentences. "I discovered we spoke the same language," he says. Two years later, when LeCun finished his PhD thesis, which explored a technique similar to backpropagation, Hinton flew to Paris and joined the thesis committee, though he still knew almost no French. Typically, when reading research papers, he skipped the math and read the text. With LeCun's thesis, he had no choice but to skip the text and read the math. For the thesis defense, the agreement was that Hinton would ask questions in English and LeCun would answer in French. This worked well enough, except that Hinton couldn't understand the answers.

After a prolonged winter, neural networks were beginning to come in from the cold. Dean Pomerleau was working on his self-driving car at Carnegie Mellon. Meanwhile, Sejnowski was making waves with something he called "NETtalk." Using a hardware device that could generate synthetic sounds—a bit like the robotic speech box used by the British physicist Stephen Hawking after a neurodegenerative disorder took his voice—Sejnowski built a neural network that could learn to read aloud. As it analyzed children's books filled with English words and their matching phonemes (how each letter was pronounced), it could pronounce other words on its own. It could learn when a "gh" was pronounced like an "f" (as in "enough") and when a "ti" was pronounced "sh" (as in "nation"). When he gave lectures at conferences, Sejnowski played a recording of the device during each stage in its training. At first, it babbled like a baby. After half a day, it began to pronounce recognizable words. After a week, it could read aloud. "Enough." "Nation." "Ghetto." "Tint." His system showed both what a neural network could do and how it worked. As Sejnowski took this creation on a tour of academic conferences—and on the *Today* show, sharing it with millions of television viewers—it helped galvanize connectionist research on both sides of the Atlantic.

After finishing his degree, LeCun followed Hinton to Toronto for a yearlong postdoc. He carried two suitcases from France—one packed with clothes, the other with his personal computer. Though the two men got on well, their interests were hardly identical. Whereas Hinton was driven by a need to understand the brain, LeCun, a trained electrical engineer, was also interested in the hardware of computing, the mathematics of neural networks, and the creation of artificial intelligence in the broadest sense of the term. His career had been inspired by the debate between Chomsky and Piaget. It had also been inspired by the Hal 9000 and other machines of the future depicted in Stanley Kubrick's *2001: A Space Odyssey*, which he saw in 70mm Cinerama at a Paris theater when he was nine years old. More than four decades later, when he built one of the world's leading corporate labs, he hung framed stills from the film on the walls. Across his career, as he explored neural networks and other algorithmic techniques, he also designed computer chips and off-road self-driving cars. "I did whatever I could get my hands on," he says. He embodied the way that artificial intelligence, an academic pursuit that was more of an attitude than a formal science, blended so many disparate forms of research, pulling them all into what was often an overly ambitious effort to build machines that behave like humans. Mimicking even a small part of human intelligence, as Hinton aimed to do, was an immense task. Applying intelligence to cars and planes and robots was harder still. But LeCun was more practical and more grounded than many of the other researchers who would later rise to the fore. In the decades to come, some voices questioned whether neural networks would ever be useful. Then, once their power was apparent, some questioned whether AI would destroy humankind. LeCun thought both questions were ridiculous, and he was never slow to say so, both in private and in public. As he said decades later, in a video shown the night he received the Turing Award, the Nobel Prize of computing: "I was definitely thinking I was right the whole time." He believed that neural networks were the path to very real and very useful technology. And that was what he said.

His breakthrough was a variation on the neural network modeled on the visual cortex, the part of the brain that handles sight. Inspired by the work of a Japanese computer scientist named Kunihiko Fukushima, he called this a "convolutional neural network." Just as different sections of the visual cortex process different sections of the light captured by your eyes, a convolutional neural network cut an image into squares and analyzed each one separately, finding small patterns in these squares and building them into larger patterns as information moved through its web of (faux) neurons. It was an idea that would define LeCun's career. "If Geoff Hinton is a fox, then Yann LeCun is a hedgehog," says University of California–Berkeley professor Jitendra Malik, borrowing a familiar analogy from the philosopher Isaiah Berlin. "Hinton is bubbling with these ideas, zillions and zillions of ideas bouncing in different directions. Yann is much more single-minded. The fox knows many little things and the hedgehog knows one big thing."

LeCun first developed his idea during his year with Hinton in Toronto. Then it blossomed when he moved to Bell Labs, which had the massive amounts of data needed to train LeCun's convolutional neural network (thousands of dead letters). It also had the extra processing power needed to analyze these envelopes (a brand-new Sun Microsystems workstation). He told his boss that he joined Bell Labs because he was promised his own workstation and wouldn't have to share a machine as he did during his postdoc in Toronto. Weeks after joining the lab, using the same basic algorithm, he had a system that could recognize handwritten digits with an accuracy that exceeded any other technology under development at AT&T. It worked so well that it soon found a commercial application. In addition to Bell Labs, AT&T owned a company called NCR that sold cash registers and other business equipment, and by the mid-'90s, NCR was selling LeCun's technology to banks as a way of automatically reading handwritten checks. At one point, LeCun's creation read more than 10 percent of all checks deposited in the United States.

But he hoped for more. Inside the glass walls of the Bell Labs complex in Holmdel—known as "the world's largest mirror"—LeCun and his colleagues also designed a microchip they called ANNA. It was an acronym inside an acronym. ANNA was the acronym for Analog Neural Network ALU, and ALU stood for Arithmetic Logic Unit, a kind of digital circuit suited to running the mathematics that drove neural networks. Instead of running their algorithms using ordinary chips built for just any task, Le-

Cun's team built a chip for this one particular task. That meant it could handle the task at speeds well beyond the standard processors of the day: about 4 billion operations a second. This fundamental concept—silicon built specifically for neural networks—would remake the worldwide chip industry, though that moment was still two decades away.

Not long after LeCun's bank scanner reached the market, AT&T, the former national telephone system that divided into so many smaller companies over the decades, divided once again. NCR and LeCun's research group were suddenly separated, and the bank scanner project was disbanded, leaving LeCun disillusioned and depressed. As his group moved toward technologies involving the World Wide Web, which was just beginning to take off in mainstream America, he stopped working on neural networks entirely. When the company started laying off researchers, LeCun made it clear he wanted a pink slip, too. "I don't give a shit what the company wants me to work on," he told the head of the lab, "I'm working on computer vision." The pink slip duly arrived.

In 1995, two Bell Labs researchers—Vladimir Vapnik and Larry Jackel—made a bet. Vapnik said that within ten years, "no one in their right mind would use neural nets." Jackel sided with the connectionists. They bet a "fancy dinner," with LeCun serving as witness when they typed up the agreement and signed their names. Almost immediately, it started to look as though Jackel would lose. As the months passed, another chill settled over the wider world of connectionist research. Pomerleau's truck could drive itself. Sejnowski's NETtalk could learn to read aloud. And LeCun's bank scanner could read handwritten checks. But it was clear the truck couldn't deal with anything more than private roads and the straight lines of a highway. NETtalk could be dismissed as a party trick. And there were other ways of reading checks.

LeCun's convolutional neural networks didn't work when analyzing more complex images, like photos of dogs, cats, and cars. It wasn't clear if they ever would. As it turned out, Jackel eventually won the bet, but it was a hollow victory. Ten years after their wager, researchers might still have been using neural nets, but the technology couldn't do that much more than it had done on LeCun's desktop machine all those years before. "I won the bet mostly because Yann didn't give up," Jackel says. "He was largely ignored by the outside community, but he didn't give up."

Not long after the bet was settled, during a lecture on artificial intelligence, a Stanford University computer science professor named Andrew Ng described neural networks to a roomful of graduate students. Then he added a caveat. "Yann LeCun," he said, "is the only one who can actually get them to work." But even LeCun was unsure of the future. With a few wistful words on his personal website, he wrote about his chip research as something stuck in the past. He described the silicon processors he helped build in New Jersey as "the first (and maybe the last) neural net chips to actually do something useful." Years later, when asked about these words, he dismissed them, quickly pointing out that he and his students had returned to the idea by the end of the decade. But the uncertainty he felt was there on the page. Neural networks did need more computing power, but no one realized just how much they needed. As Geoff Hinton later put it: "No one ever thought to ask: 'Suppose we need a million times more?'"

WHILE Yann LeCun was building his bank scanner in New Jersey, Chris Brockett was teaching Japanese in the Department of Asian Languages and Literature at the University of Washington. Then Microsoft hired him as an AI researcher. The year was 1996, not long after the tech giant created its first dedicated research lab.

Microsoft aimed to build systems that could understand natural language—the everyday way people write and talk. At the time, this was the work of linguists. Language experts like Brockett, who had studied linguistics and literature in his native New Zealand and later in Japan and the U.S., spent their days writing detailed rules meant to show machines how humans pieced their words together. They would explain why "time flies," carefully separate "contract" the noun from "contract" the verb, describe in minute detail the strange and largely unconscious way that English speakers choose the order of their adjectives, and so on. It was a task reminiscent of the old Cyc project in Austin or the driverless car work at Carnegie Mellon before Dean Pomerleau came along—an effort to re-create human knowledge that wouldn't reach its endgame for decades, no matter how many linguists Microsoft hired. In the late '90s, following the lead of prominent researchers like Marvin Minsky and John McCarthy, this is how most universities and tech companies built computer vision and speech recognition as well as natural language understanding. Experts pieced the technology together one rule at a time.

Sitting in an office at Microsoft headquarters just outside Seattle, Brockett spent nearly seven years writing the rules of natural language. Then, one afternoon in 2003, inside an airy conference room down the hall, two of his colleagues unveiled a new project. They were building a system that translated between languages using a technique based on statistics—how often each word appeared in each language. If a set of words appeared with the same frequency and the same context in both languages, that was the likely translation. The two researchers had started the project only six weeks earlier, and it was already producing results that looked at least a little like real language. As he watched the presentation, sitting at the back of the crowded room, perched atop a long row of trash cans, Brockett had a panic attack—which he thought was a heart attack—and was rushed to the hospital. He later called it his "come-to-Jesus moment," when he realized he had spent six years writing rules that were now obsolete. "My fifty-two-year-old body had one of those moments when I saw a future where I wasn't involved," he says.

The world's natural language researchers soon overhauled their approach, embracing the kind of statistical models unveiled that afternoon at the lab outside Seattle. This was just one of many mathematical methods that spread across the larger community of AI researchers in the 1990s and on into the 2000s, with names like "random forests," "boosted trees," and "support vector machines." Researchers applied some to natural language understanding, others to speech recognition and image recognition. As the progress of neural networks stagnated, many of these other methods matured and improved and came to dominate their particular corners of the AI landscape. They were all a (very) long way from perfection. Though the early success of statistical translation had been enough to send Chris Brockett to the hospital, it worked only to a point, and only when applied to short phrases—pieces of a sentence. Once a phrase was translated, a complex set of rules was needed to get it into the right tense and apply the right word endings and line it up with all the other short phrases in a sentence. Even then, the translation was jumbled and only vaguely correct, like that childhood game where you build a story by rearranging little slips of paper holding just a handful of words. But this was still beyond what a neural network could do. By 2004, a neural network was seen as the third best way to tackle any task—an old technology whose best days were behind it. As one researcher told Alex Graves, then a young graduate student studying neural networks in Switzerland: "Neural networks are for people who don't understand stats." While hunting for a major at Stanford, a nineteen-year-old undergraduate named Ian Goodfellow took a class in what was called cognitive science—the study of thought and learning—and at one point, the lecturer dismissed neural networks as a technology that couldn't handle "exclusive-or." It was a forty-year-old criticism debunked twenty years earlier.

In the United States, connectionist research nearly vanished from the top universities. The one serious lab was at New York University, where Yann LeCun took a professorship in 2003, his hair pulled back in a ponytail. Canada became a haven for those who still believed in these ideas. Hinton was in Toronto, and one of LeCun's old colleagues from Bell Labs, Yoshua Bengio, another Paris-born researcher, oversaw a lab at the University of Montreal. During these years, Ian Goodfellow applied to graduate schools in computer science, and several offered him a spot, including Stanford, Berkeley, and Montreal. He preferred Montreal, but when he visited, a Montreal student tried to talk him out of it. Stanford was the number-three-ranked computer science program in North America. Berkeley was number four. And both were in sunny California. The University of Montreal was ranked somewhere around a hundred fifty, and it was cold.

"Stanford! One of the most prestigious universities in the world!" this Montreal student told him as they walked through the city in late spring, snow still on the ground. "What the hell are you thinking?"

"I want to study neural networks," Goodfellow said.

The irony was that as Goodfellow explored neural networks in Montreal, one of his old professors, Andrew Ng, after seeing the research that continued to emerge from Canada, was embracing the idea in his lab at Stanford. But he was very much an outlier, both at his own university and across the wider community, and he didn't have the data needed to convince those around him that neural networks were worth exploring. During these years, he gave a presentation at a workshop in Boston that trumpeted neural networks as the wave of the future. In the middle of his talk, the Berkeley professor Jitendra Malik, one of the de facto leaders of the computer vision community, stood up, Minsky-like, and told him this was nonsense, that he was making his self-satisfied claim with absolutely no evidence to support it.

Around the same time, Hinton submitted a paper to NIPS, the conference where he would later auction off his company. This was a conference conceived in the late 1980s as a way for researchers to explore neural networks of all kinds, the biological as well as the artificial. But the conference organizers rejected Hinton's paper because they had accepted another neural network paper and thought it would be unseemly to accept two in the same year. "Neural" was a bad word, even at a conference dedicated to Neural Information Processing Systems. Across the field, neural networks showed up in less than 5 percent of all published research papers. When submitting papers to conferences and journals, hoping to improve their chances of success, some researchers would replace the words "neural network" with very different language, like "function approximation" or "nonlinear regression." Yann LeCun removed the word "neural" from the name of his most important invention. "Convolutional neural networks" became "convolutional networks."

Still, papers that LeCun viewed as undeniably important were rejected by the AI establishment, and when they were, he could be openly combative, adamant that his views were the right views. Some saw this as unfettered confidence. Others believed it betrayed an insecurity, an underlying remorse that his work wasn't recognized by the leaders of the field. One year, Clément Farabet, one of his PhD students, built a neural network that could analyze a video and separate different kinds of objects—the trees from the buildings, the cars from the people. It was a step toward computer vision for robots or self-driving cars, able to perform its task with fewer errors than other methods and at faster speeds, but reviewers at one of the leading vision conferences summarily rejected the paper. LeCun responded with a letter to the conference chair saying that the reviews were so ridiculous, he didn't know how to begin writing a rebuttal without insulting the reviewers. The conference chair posted the letter online for all to see, and though he removed LeCun's name, it was obvious who had written it.

The only other places to really study neural networks were in Europe or Japan. One was a lab in Switzerland overseen by a man named Jürgen Schmidhuber. As a child, Schmidhuber had told his younger brother that the human brain could be rebuilt with copper wires, and from the age of fifteen, his ambition was to build a machine more intelligent than he was—then retire. He embraced neural networks as an undergraduate in the 1980s, and then found, as he emerged from graduate school, that his ambitions interlocked with those of an Italian liqueur magnate named Angelo Dalle Molle. At the end of the decade, after building a fortune with a liqueur made from artichokes, Dalle Molle erected an AI lab on the shores of Lake Lugano in Switzerland, near the Italian border, intent on overhauling society with intelligent machines that could handle all the work that traditionally fell to humans. Soon the lab hired Schmidhuber.

He was six feet two inches tall with a trim build and a square jaw. He dressed in fedoras, driving caps, and Nehru jackets. "You could imagine him stroking a white cat," one of his former students says, nodding to the villain of the early James Bond films, Ernst Blofeld, who wore his own Nehru jackets. Schmidhuber's attire somehow matched the lab in Switzerland, which also looked like something out of a Bond film—a fortress beside a European lake, surrounded by palm trees. Inside the Dalle Molle Institute for Artificial Intelligence Research, Schmidhuber and one of his students developed what they described as a neural network with short-term memory. It could "remember" data it had recently analyzed and, using this recall, improve its analysis each step along the way. They called it an LSTM, for Long Short-Term Memory. It didn't actually do much, but Schmidhuber believed this kind of technology would deliver intelligence in the years to come. He described some neural networks as having not just memory but sentience. "We have consciousness running in our lab," he would say. As one student later put it, with some affection: "He sounded like some kind of a maniac."

Hinton would joke that LSTM stood for "looks silly to me." Schmidhuber was a particularly colorful example of a long tradition among AI researchers, beginning with Rosenblatt, Minsky, and McCarthy. Since the field was created, its leading figures had casually promised lifelike technology that was nowhere close to actually working. Sometimes, this was a way of raising money, either from government agencies or venture capitalists. Other times, it was a genuine belief that artificial intelligence was nigh. This kind of attitude could drive research forward. Or, if the technology didn't live up to the hype, it could stall progress for years.

The connectionist community was tiny. Its leaders were European—English, French, German. Even the political, religious, and cultural beliefs harbored by these researchers sat outside the American mainstream. Hinton was an avowed socialist. Bengio gave up his French citizenship because he didn't want to serve in the military. LeCun called himself a "militant atheist." Though this was not the kind of language Hinton would ever use, he felt much the same. He often recalled a moment from his teenage years when he was sitting in the chapel at his English public school, Clifton College, listening to a sermon. From the pulpit, the speaker said that in Communist countries, people were forced to attend ideological meetings and weren't allowed to leave. Hinton thought: "That is exactly the situation I am in." He would retain these very personal beliefs—atheism, socialism, connectionism—in the decades to come, though after selling his company to Google for $44 million, he enjoyed calling himself "gauche caviar." "Is that the right term?" he would ask, knowing full well that it was.

AS difficult as the '90s were for LeCun, they were more difficult for Hinton. Not long after he moved to Toronto, he and his wife adopted two children from South America, a boy, Thomas, from Peru and a girl, Emma, from Guatemala. Both were under the age of six when his wife felt a pain in her abdomen and started losing weight. Though this continued for several months, she declined to see a doctor, holding tight to her belief in homeopathic medicine. When she finally relented, she was diagnosed with ovarian cancer. Even then, she insisted on treating it homeopathically, not with chemotherapy. She died six months later.

Hinton thought his days as a researcher were over. He had to care for his children, and Thomas, who had what the family called "special needs," required extra attention. "I was used to spending my time thinking," Hinton says. Two decades later, when he accepted the Turing Award alongside LeCun, he thanked his second wife, a British art historian named Jackie Ford, for saving his career when they married in the late '90s and she helped raise his children. They had met years earlier at the University of Sussex and dated for a year in England before splitting up when he moved to

San Diego. When they reunited, he moved to Britain and took a job at University College London, but they soon returned to Canada. He felt his children were more welcome in Toronto.

So, at the turn of the millennium, Hinton was back in his office in the corner of the computer science building at the University of Toronto, looking out on the cobblestone street that ran through the heart of the campus. The windows were so large, they sucked the warmth from his office, radiating heat out into the subzero cold outside. This office became the hub of the small community of researchers who still believed in neural networks, in part because of Hinton's history in the field, but also because his creativity, enthusiasm, and wry sense of humor drew people to him, even in small moments. If you sent an email asking if he preferred to be called Geoffrey or Geoff, his response was equal parts clever and endearing:

> I prefer Geoffrey.
>
> Thanks,
> Geoff

A researcher named Aapo Hyvärinen once published an academic paper with an acknowledgment that summed up both Hinton's sense of humor and his belief in ideas over mathematics:

> *The basic idea in this paper was developed in discussions with Geoffrey Hinton, who, however, preferred not to be a coauthor because the paper contained too many equations.*

He rated ideas on how much weight he had lost because he had forgotten to eat. One student said that the best Christmas present

Hinton's family could ever give was an agreement to let him go back into the lab and do a little more research. And, as many colleagues often said, he had a lifelong habit of running into a room, saying he finally understood how the brain worked, explaining his new theory, leaving just as quickly as he came, and then returning days later to say that his theory about the brain was all wrong but he now had a new one.

Russ Salakhutdinov, who would become one of the world's leading connectionist researchers and a seminal hire at Apple, had quit the field by the time he ran into Hinton at the University of Toronto in 2004. Hinton told him about a new project, a way of training massive neural networks one layer at a time, feeding them far more data than was possible in the past. He called them "deep belief networks," and in that moment, he coaxed Salakhutdinov back into the fold. It was the name, as much as anything, that pulled him in. A young student named Navdeep Jaitly was drawn to the Toronto lab after visiting a professor down the hall and seeing just how many students were lined up outside Hinton's office. Another student, George Dahl, noticed a similar effect across the wider world of machine learning research. Every time he identified an important research paper—or an important researcher—there was a direct connection to Hinton. "I don't know whether Geoff picks people that end up being successful or he somehow makes them successful. Having experienced it, I think it's the latter," Dahl says.

The son of an English professor, Dahl was an academic idealist who compared joining a graduate school to entering a monastery. "You want to have an inescapable destiny, some sort of calling that will see you through the dark times when your faith lapses," he liked to say. His calling, he decided, was Geoff Hinton. He was not alone. Dahl had visited another machine learning group at the

University of Alberta, where a student named Vlad Mnih tried to convince him this was where he belonged, not Toronto. But when Dahl showed up at the University of Toronto that fall and walked into the converted supply closet where the university assigned him a desk, Mnih was there, too. He had moved to Hinton's lab over the summer.

In 2004, as interest in neural networks waned across the field, Hinton doubled down on the idea, hoping to accelerate the research inside this small community of connectionists. "The theme in Geoff's group was always: What's old is new," Dahl says. "If it's a good idea, you keep trying for twenty years. If it's a good idea, you keep trying it until it works. It doesn't stop being a good idea because it doesn't work the first time you try it." With a modicum of funding from the Canadian Institute for Advanced Research—less than $400,000 a year—Hinton created a new collective focused on what he called "neural computation and adaptive perception," hosting twice-yearly workshops for the researchers who still held these connectionist beliefs, including computer scientists, electrical engineers, neuroscientists, and psychologists. LeCun and Bengio were part of this effort, and so was Kai Yu, the Chinese researcher who would one day join Baidu. Hinton later compared this collective to Bob Woodward and Carl Bernstein working together—not separately—as they chased the Watergate story. It provided a way of sharing ideas. In Toronto, one idea was a new name for this very old technology.

When Hinton gave a lecture at the annual NIPS conference, then held in Vancouver, on his sixtieth birthday, the phrase "deep learning" appeared in the title for the first time. It was a cunning piece of rebranding. Referring to the multiple layers of neural networks, there was nothing new about "deep learning." But it was an evocative term designed to galvanize research in an area that had once again fallen from favor. He knew the name was a good one when, in the middle of the lecture, he said everyone else was doing "shallow learning," and his audience let out a laugh. In the long term, it would prove to be a masterly choice. And it immediately fed the reputation of this tiny group of researchers working on the fringes of academia. One year at NIPS, someone put together a spoof video in which one person after another embraced deep learning as if it were Scientology or the Peoples Temple.

"I used to be a rock star," said one convert. "Then I found deep learning."

"Hinton is the leader," said another. "Follow the leader."

It was funny because it was true. This was a decades-old technology that had never quite proven its worth. But some still believed.

Fifty years after the summer conference that launched the AI movement, Marvin Minsky and many of the other founding fathers returned to Dartmouth for an anniversary celebration. This time, Minsky was onstage, and another researcher stood up in the audience. It was Terry Sejnowski, now a professor at the Salk Institute, after moving from Baltimore in the East to San Diego in the West. Sejnowski told Minsky that some AI researchers saw him as the devil because he and his book had halted the progress of neural networks.

"Are you the devil?" Sejnowski asked. Minsky waved the question aside, explaining the many limitations of neural networks and pointing out, rightly, that they had never done what they were supposed to do.

So Sejnowski asked again: "Are you the devil?"

Exasperated, Minsky finally answered: "Yes, I am."

4

BREAKTHROUGH

"DO WHAT YOU WANT AT GOOGLE—NOT WHAT GOOGLE WANTS YOU TO DO."

On December 11, 2008, Li Deng walked into a hotel in Whistler, British Columbia, just north of Vancouver, at the foot of the snow-covered peaks that would soon host the ski races for the 2010 Winter Olympics. He wasn't there for the skiing. He was there for the science. Each year, hundreds of researchers traveled to Vancouver for NIPS, the annual AI conference, and when it was done, most made the short trip to Whistler for the more intimate NIPS "workshops," two days of academic presentations, Socratic debate, and hallway conversation that probed the near future of artificial intelligence. Born in China and educated in the United States, Deng had spent his career building software that aimed to recognize spoken words, first as a professor at the University of Waterloo in Canada, then as a researcher at Microsoft's central R&D lab near Seattle. Companies like Microsoft had been selling "speech rec" software for more than a decade, billing the technology as a way of taking automatic dictation on PCs and laptops, but the undeniable truth was that it didn't work that well, getting more words wrong than right as you enunciated into a long-necked desktop microphone. Like most AI research at the time, the technology improved at a glacial pace. At Microsoft, Deng and his team had spent three years building their latest speech system, and it was maybe 5 percent more accurate than the last one. Then, one evening in Whistler, he ran into Geoff Hinton.

He knew Hinton from his time in Canada. In the early '90s, during the brief renaissance in connectionist research, one of Deng's students wrote a thesis exploring neural networks as a way of recognizing speech, and Hinton, by then a professor in Toronto, joined the thesis committee. The two researchers saw little of each other in the years that followed, as connectionism fell from favor across both industry and academia. Though Hinton held tight to the idea of a neural network, speech recognition was never more than a side interest at his lab in Toronto, and that meant he and Deng moved in very different circles. But when they walked into the same room at the Hilton Whistler Resort and Spa—an all but empty room where a few researchers sat at tables waiting for someone to ask them questions about their latest research—Deng and Hinton fell straight into conversation. Deng, enormously excitable and even more talkative, fell straight into conversation with just about everyone. Hinton had an agenda.

"What's new?" Deng asked.

"Deep learning," Hinton answered. Neural networks, he said, were beginning to work with speech.

Deng didn't really believe it. Hinton was not a speech researcher, and neural networks had never quite worked with *anything*. At Microsoft, Deng was developing a new speech method of his own, and he didn't really have time for another trip into the algorithmic unknown. But Hinton was insistent. His research wasn't receiving much attention, he said, but over the past few years, he and his students had published a series of papers detailing his deep belief networks, which could learn from much larger amounts of data than previous techniques and were now approaching the performance of the leading speech methods. "You must try them," Hinton kept saying. Deng said he would, and they exchanged email addresses. Then months went by.

In the summer, with a little extra time on his hands, Deng started reading the literature on what was then called neural speech recognition. He was impressed enough with its performance that he emailed Hinton, suggesting they organize a new Whistler workshop around the idea, but he still questioned the long-term prospects of a technique that was systematically ignored by the worldwide speech community. It worked well on simple tests, but so did any number of other algorithmic methods. Then, as the next Whistler gathering approached, Hinton sent Deng another email, attaching an early draft of a research paper that pushed his techniques even further. It showed that after analyzing about three hours of spoken words, a neural network could match the performance of even the best speech methods. Deng still didn't believe it. The Toronto researchers described their technology in a way that was terribly hard to understand, and they'd tested their system on a database of sounds recorded in a lab, not real-world speech. Hinton and his students were treading into an area of research they weren't completely familiar with, and it showed. "The paper just wasn't done properly," Li Deng says. "I just couldn't believe they had gotten the same results as me." So he asked to see the raw data from their tests. When he opened the email and looked at the data and saw for himself what the technology could do, he believed.

THAT summer, Li Deng invited Hinton to spend some time at Microsoft's research lab in Redmond, Washington. Hinton agreed. But first he had to get there. In recent years, his back problems had progressed to the point that he questioned, once again, whether his research could continue. He had slipped a disk forty years earlier while trying to move a brick-filled storage heater for his mother, and it became increasingly unstable as the years passed. These days it could slip when he leaned over or just sat down. "It was a combination of genetics, stupidity, and bad luck, like everything else that goes wrong in life," he says. The only solution, he decided, was to stop sitting down (except for, as he puts it, "a couple of minutes at a time, once or twice per day," thanks to the inevitability of "biology"). At his lab in Toronto, he would meet with students while lying flat on his office desk or across a cot he kept against the wall, trying to ease the pain. It also meant he couldn't drive and he couldn't fly.

So, in the fall of 2009, he took the subway to the bus station in downtown Toronto, got in line early so he could claim the back seat of the bus to Buffalo, lay down, and pretended to be asleep so that no one would try to move him. "In Canada, that works," he says. (On the return trip, from the U.S. back into Canada, it did not: "I laid down in the back seat and pretended to be asleep and this guy came up and kicked me.") When he arrived in Buffalo, he arranged the visa he needed to work in the lab at Microsoft, then took the nearly three-day train ride across the country to Seattle. Deng didn't realize Hinton's back was an issue until he heard how long the trip would take. Before the train arrived, he ordered a standing desk for his office, so they could work side by side.

Hinton arrived in mid-November, lying down in the back seat of a taxi for the ride across the floating bridge that spanned Lake

Washington, connecting Seattle to its Eastside and then Redmond, a small suburban town dominated by the many medium-sized office buildings of a very large corporation. He joined Deng in an office on the third floor of Microsoft Building 99, a granite-and-glass building that served as the heart of the company's R&D lab. This was the same lab where the linguist Chris Brockett had had his panic attack—an academic-style lab that focused not on markets and money like the rest of Microsoft but on the technologies of the future. When it opened in 1991, just as Microsoft came to dominate the international software market, one of the lab's primary goals was technology that could recognize spoken words, and over the next fifteen years, paying unusually high salaries, the company hired many of the top researchers in the field, including Li Deng. But by the time Hinton arrived in Redmond, Microsoft's place in the world was changing. The balance of power was shifting away from the software giant and into other areas of the tech landscape. Google, Apple, Amazon, and Facebook were on the rise, taking hold of new markets and new money—Internet search, smartphones, online retail, social networking. Microsoft still dominated computer software with its Windows operating system, which ran most desktop PCs and laptops, but after expanding into one of the world's largest companies—and taking on all the usual corporate bureaucracy—it was slow to change direction.

Inside Building 99, where four stories of labs, conference rooms, and offices encircled a large atrium and a small coffee shop, Hinton and Deng planned to build a prototype based on the research in Toronto, training a neural network to recognize spoken words. This was no more than a two-person project, but they had trouble getting down to work. Hinton needed a password to log in to Microsoft's computer network, and the only way to get a password was over a company phone, which required its own password. They sent countless email messages trying to get a password for a phone, and when that didn't work, Deng walked Hinton up to the tech-support desk on the fourth floor. Microsoft had a special rule that allowed a temporary network password if someone was visiting for only a day, and the woman sitting at the desk gave them one. But when Hinton asked if it would still work the next morning, she took it back. "If you're staying more than a day," she said, "you can't have it."

Once they finally found a way onto the network, the project came together in just a few days. At one point, as Hinton was typing computer code into his desktop machine, Deng started typing alongside him, on the same keyboard. It was typical of the ever-excitable Deng, and Hinton had never seen anything like it. "I was used to people interrupting each other," he says. "I was not used to being interrupted while typing code on a keyboard by someone else typing code on the same keyboard." They built their prototype using a programming language called MATLAB, and it spanned no more than ten pages of code, written mostly by Hinton. As much as Hinton downplayed his skills as a mathematician and computer scientist, Deng was struck by the elegant simplicity of his code. "It's so clear," Deng thought. "Line by line." But it wasn't just the clarity that impressed him. Once they trained the system on Microsoft's speech data, *it worked*—not as well as the leading systems of the day, but well enough to cement Deng's realization that this was the near future of speech recognition. Commercial systems used other, handcrafted methods to recognize speech, and they didn't really work. But Deng could see that he and Hinton had created a system that could grow more powerful as it learned from larger amounts of data.

What their prototype still lacked was the extra processing power needed to analyze all that data. In Toronto, Hinton made use of a very particular kind of computer chip called a GPU, or graphics processing unit. Silicon Valley chip makers like Nvidia originally designed these chips as a way of quickly rendering graphics for popular video games like Halo and Grand Theft Auto, but somewhere along the way, deep learning researchers realized GPUs were equally adept at running the math that underpinned neural networks. In 2005, three engineers had tinkered with the idea inside the same Microsoft lab where Deng and Hinton would later build their speech prototype, and a team at Stanford University stumbled onto the same technical trick around the same time. These chips allowed neural nets to learn from more data in less time—an echo of Yann LeCun's work at Bell Labs in the early '90s. The difference was that GPUs were off-the-shelf hardware. Researchers didn't have to build new chips to accelerate the progress of deep learning. Thanks to games like Grand Theft Auto and gaming consoles like the Xbox, they could use chips that were already there for the taking. In Toronto, Hinton and his two students Abdel-rahman Mohamed and George Dahl, the English professor's son, trained their speech system using these specialized chips, and this is what pushed it beyond the state of the art.

After Hinton finished his brief stay at Microsoft, Deng insisted that both Mohamed and Dahl visit Building 99, and he wanted them there at separate times so the project would continue unabated for the next several months. Agreeing to this extended experiment, Hinton and his students explained that the project wouldn't succeed without a very different breed of hardware, including a $10,000 GPU card. At first, Deng balked at the price. His boss, Alex Acero, who would one day oversee Siri, the digital assistant on the Apple iPhone, told him this was an unnecessary expense. GPUs were for games, not AI research. "Don't waste your money," he said, telling Deng to skip the expensive Nvidia gear and buy generic cards at the local Fry's Electronics store. But Hinton urged Deng to push back, explaining that cheap hardware would defeat the purpose of the experiment. The idea was that a neural network would analyze Microsoft's speech data over the course of several days, and those generic cards might flame out if they ran that hot for that long. But his larger point was that neural networks thrived on extra processing power. Deng needed to buy not only a $10,000 GPU card but maybe more than one, plus a specialized server that could run the card. That would cost just as much as the card. "It would cost you about $10,000," Hinton said in an email to Deng. "We are about to order 3, but then we are a well funded Canadian university, not a poor software seller." In the end, Deng bought the necessary hardware.

That year, Microsoft hired a man named Peter Lee as the new head of its Redmond research lab. A trained researcher with the air of an administrator, Lee had spent more than two decades at Carnegie Mellon, rising to become the head of the computer science department. When he first joined Microsoft and began reviewing the lab's research budget, he stumbled onto a worksheet that listed expenses for Deng's speech project, including payments to Hinton, Mohamed, and Dahl; funds for the speech workshop in Whistler; and charges for the GPUs. Lee was gobsmacked. He thought the whole arrangement was one of the stupidest ideas to ever cross his desk. He had known Hinton at Carnegie Mellon in the '80s, and he thought neural networks were ridiculous then. Now he thought they were madness. But by the time he arrived in Redmond, the project was already in motion. "I sometimes think that if I had been hired at Microsoft a year earlier," Lee says, "none of this stuff would have happened."

The breakthrough came that summer, when George Dahl visited the lab. A tall man with big features and small glasses, Dahl had embraced machine learning research as a lifelong pursuit during his sophomore year in college, seeing it as an alternative breed of computer programming—something that would let you tackle a problem even if you didn't completely know how to tackle it. You could just let the machine learn. He was steeped in neural networks, but he wasn't really a speech researcher. "The only reason I started working on speech was because everyone else in Geoff's group was working on vision," he often said. He aimed to show that the ideas brewing inside Hinton's lab would work with more than just images. And he did. "George didn't know much about speech," Li Deng says, "but he knew GPUs." At Microsoft, using those $10,000 cards to train a neural network on spoken words collected by the company's Bing voice search service, Dahl pushed Hinton's speech prototype beyond the performance of anything else under development at the company. What he and Mohamed and Hinton showed was that a neural network could sift through a sea of very noisy speech and somehow find the stuff that mattered, the patterns that no human engineer could ever pinpoint on their own, the telltale signs that distinguished one subtle sound from another, one word from another. It was an inflection point in the long history of artificial intelligence. In a matter of months, a professor and his two graduate students matched a system that one of the world's largest companies had worked on for more than a decade. "He is a genius," Deng says. "He knows how to create one impact after another."

A few months later, standing at his desk in Toronto, looking out over the cobblestones of the King's College Road, Geoff Hinton opened an email from a man he'd never heard of. His name was Will Neveitt, and he asked if Hinton could send a student to Google headquarters in Northern California. With their speech recognition work, Hinton and his students had sparked a chain reaction across the tech industry. After seeding a new speech project at Microsoft—and publishing their research for all to see—they repeated the trick at a second tech giant: IBM. In the fall of 2010, nine months after visiting Microsoft, Abdel-rahman Mohamed began working with IBM's Thomas J. Watson Research Center, another majestic Eero Saarinen creation, complete with mirrored windows, this one tucked among the rolling hills just north of New York City. Now it was Google's turn.

Mohamed was still working with IBM, and George Dahl was busy with other research, so Hinton turned to a student who had almost nothing to do with their speech work. This was Navdeep Jaitly, the son of Indian immigrants to Canada, who had only recently embraced AI research after several years as a computational biologist. A particularly congenial researcher with a closely shaved head, he worked alongside Dahl in the converted supply closet down the hall from Hinton, and he was in the market for an industry internship. Hinton had tried to find him a spot at Research in Motion (RIM), maker of the BlackBerry smartphone, but the Canadian company said it wasn't interested in speech recognition. RIM, whose keyboard-equipped devices dominated the phone market just a few years earlier, had already missed the leap to touchscreen smartphones. Now the next big leap was about to pass the company by. When Hinton first offered Jaitly the Google job, he turned it down. He and his wife were about to have a baby, and because he'd already applied for a green card in the United States, he knew he couldn't get the visa he would need to actually do work at Google. But a few days later, he reconsidered, asking Will Neveitt, the Googler who'd emailed Hinton, to buy a machine filled with GPUs.

By the time his Google internship was set to begin, Neveitt had left the company. His replacement, a French-born engineer named Vincent Vanhoucke, found himself with a massive GPU machine he didn't quite know what to do with and a Canadian intern who did know but wasn't allowed to work at the office where the machine was sitting because he didn't have a visa. So Vanhoucke phoned someone in the small office Google had opened in Montreal and found an empty desk. This is where Jaitly worked that summer, almost entirely on his own, tapping into that massive GPU machine over the Internet. But first he took a brief trip to Northern California so he could meet Vanhoucke and get the GPU machine up and running. "No one else knew how to deal with that thing," Vanhoucke says. "So he had to do it."

When Jaitly arrived, the machine was tucked into a corner down the hall from Vanhoucke and the rest of the speech team. "It's humming over there behind the printer," Vanhoucke said. He didn't want it inside someone's office or anywhere close to where anyone was working. Each GPU card was equipped with a fan that whirred incessantly as it struggled to keep the hardware from overheating, and he worried that someone would get fed up with the noise and shut the machine down without realizing what it was for. He put it behind the printer so that anyone who heard the whirring of the fans would blame the printer for all the noise. This kind of machine was an oddity at Google, as it was at Microsoft, but for different reasons. In building its empire of online services, Google had erected a worldwide network of data centers that spanned hundreds of thousands of computers. Company engineers could instantly access this vast array of computing power from any Google PC or laptop. That was how they built and tested new software—not with a machine stuck in the corner behind a printer. "The culture was: Everybody runs their software in the big data centers," Vanhoucke says. "We had lots of computers, so why would you go and *buy your own computer*?" The rub was the machines in Google's data centers didn't include GPU chips, and that's what Jaitly needed.

He wanted to do what Mohamed and Dahl had done at Microsoft and IBM: rebuild the company's existing speech system with a neural network. But he also wanted to go further. Parts of the systems at Microsoft and IBM still leaned on other technologies, and Jaitly aimed to expand what the neural network learned, hoping to eventually build a system that learned *everything* by analyzing spoken words. Before Jaitly left Toronto, Dahl, ever the academic, told him not to listen to the big corporation. "Do what you want at Google—not what Google wants you to do," he said. So when Jaitly met with Vanhoucke and others in California, he proposed a bigger neural network. At first, they balked. Even training a smaller neural network took days, and if Jaitly trained a network on Google's data, it could take weeks. He was there only for the summer. One Googler asked Jaitly if he could train a network on two thousand hours of spoken words—and then Jaitly balked. In Toronto, Mohamed and Dahl had trained on three hours of data. At Microsoft, they used twelve. Google was a place where all the data was bigger, as the company vacuumed up text, sounds, and video through its massively popular online services, including everything from Google Search to YouTube. But Jaitly held his ground, and when the meeting ended, he emailed Hinton.

"Has anyone ever trained on two thousand hours?" he asked.

"No," Hinton said. "But I don't see why it wouldn't work."

After traveling to Montreal and tapping into that whirring GPU machine over the Internet, Jaitly trained his first neural network in less than a week. When he tested this new system, it identified the wrong word about 21 percent of the time—and that was a remarkable feat. The Google speech recognition service running on the world's Android smartphones was stuck at 23 percent. After another two weeks, he pushed his system's error rate down to 18 percent. Before Jaitly started the tests, Vanhoucke and his team saw the project as an interesting experiment. They never thought it would come anywhere close to matching what Google had already built. "We just thought we were in a different league," Vanhoucke says. "It turns out we were not."

The system worked so well so quickly, Jaitly trained a second that could search YouTube videos for particular spoken words. (If you asked it to find the word "surprise," it would pinpoint the moments in the video when that word was spoken.) Google had already built a service that did the same thing, but it grabbed the wrong word 53 percent of the time. Before the summer was out, Jaitly lowered the error rate on his system to 48 percent. And he did all this almost entirely by himself. It was a blessing to be in Montreal, he thought, because no one was there to rein him in. He forgot his own limits. He worked each evening until eleven o'clock or midnight. When he got home, his wife would hand him the baby, who was up most of the night with colic. But it wasn't hard to repeat the same cycle the next day. "It was addictive," he says. "The results kept getting better."

After Jaitly and his family returned to Toronto, Vanhoucke moved his entire team onto the project. Google knew that Microsoft and IBM were building similar technology, and it wanted to get there first. The trouble was that Jaitly's system was ten times too slow to handle live queries over the Internet. No one would use it at that speed. As the team started trimming the fat, a second team joined in from a completely different part of the company. It so happened that while Jaitly was toiling away on his project in Montreal, a few other researchers, including another Hinton protégé, were building a dedicated deep learning lab back at company headquarters in California. Working alongside Vanhouke's team, this new lab pushed the technology onto Android smartphones in less than six months. At first Google didn't tell the world its speech recognition service had changed, and soon after it went live, Vanhoucke received a phone call from a small company that supplied a chip for the latest Android phones. This chip was supposed to remove background noise when you barked commands into your phone—a way of cleaning up the sound so that the speech system could more easily identify what was being said. But the company told Vanhoucke its chip had stopped working. It was no longer boosting the performance of the speech recognition service. As Vanhoucke listened to what this company was saying, it didn't take long to realize what had happened.

The new speech recognition system was so good, it made the noise-cancellation chip obsolete. In fact, it was particularly effective when the chip *couldn't* clean up the sound. Google's neural network had learned to deal with the noise.

5

TESTAMENT

"THE SPEED OF LIGHT IN A VACUUM USED TO BE ABOUT 35 MPH. THEN JEFF DEAN SPENT A WEEKEND OPTIMIZING PHYSICS."

Andrew Ng sat in a Japanese restaurant, just down the street from Google headquarters, waiting for Larry Page. Google's founder and chief executive officer was late, as Ng knew he would be. It was the end of 2010, and in recent years, Google had grown into the most powerful force on the Internet, expanding from a small but prodigiously profitable online search company into a tech empire that dominated everything from personal email to online video to smartphones. Ng, a professor of computer science at nearby Stanford University, was sitting at a table against the wall. He felt Page was less likely to be recognized and accosted on the edges of the restaurant than he would be in the middle. As he waited, one of his Stanford colleagues, Sebastian Thrun, sat beside him. Thrun was on leave from the university after Page tapped him to run a project that had only recently been revealed to the rest of the world: the

Google self-driving car. Now, with Thrun as the go-between, Ng was pitching Page a new idea.

The thirty-four-year-old Ng, a tall man who spoke in something close to a whisper, had prepared a line graph on his laptop as a way of explaining his idea, but when Page finally arrived and sat down, Ng decided that pulling a laptop from his bag during lunch with the Google CEO was somehow the wrong thing to do. So he described the idea with his hands. The line graph, he showed, went up and to the right. As a neural network analyzed more and more data, it grew more and more accurate, whether it was learning sights, sounds, or language. And what Google had was data—years of photos, videos, voice commands, and text gathered through services like Google Search, Gmail, and YouTube. He was already exploring deep learning at his lab at Stanford. Now he hoped to put Google's weight behind the idea. Thrun was building a self-driving car inside the new "moonshot" lab dubbed Google X. The two men envisioned another moonshot built on deep learning.

Born in London and raised in Singapore, Ng was the son of a Hong Kong–born medical doctor. He studied computer science, economics, and statistics at Carnegie Mellon, MIT, and the University of California–Berkeley before moving to Stanford, where his first big project was autonomous helicopters. He soon married another roboticist, announcing their engagement in the pages of the engineering magazine *IEEE Spectrum*, complete with color photos. Though he had once told a roomful of students that Yann LeCun was the only person on Earth who could coax something useful from a neural network, he moved with the tide as he saw it turn. "He was one of the few people doing other work who switched over to neural nets because he realized what was happening," Hinton says. "His PhD advisor thought he was a traitor." After requesting an invitation, he had joined the small research collective Hinton had created with funds from the Canadian government in pursuit of "neural computation." It was no coincidence that as Hinton was pushing the technology into one part of Google, Ng was moving it into another. Viewing the technology from the same vantage point, he, too, saw where it was headed. But in pitching the idea to Larry Page, he gave it an extra flourish.

As much as he was shaped by the work of Geoff Hinton, he was also the product of a 2004 book titled *On Intelligence*, written by a Silicon Valley engineer, entrepreneur, and self-taught neuroscientist named Jeff Hawkins. Hawkins invented the PalmPilot, a forerunner of the iPhone from the 1990s, but what he really wanted to do was study the brain. In his book he argued that the whole of the neocortex—the part of the brain that handled sight, hearing, speech, and reason—is driven by a single biological algorithm. If scientists could re-create this algorithm, he said, they could re-create the brain. Ng took this to heart. In his graduate lectures at Stanford, he would describe an experiment involving the brain of a ferret. If its optic nerve was disconnected from the visual cortex (where the brain handled vision) and then attached to the auditory cortex (where the brain handled hearing), the ferret could still see. As Ng explained it, these two parts of the brain used the same fundamental algorithm, and this single algorithm could be re-created in a machine. The rise of deep learning, he would say, was a movement in this direction. "Students used to come to my office and say they wanted to work on building intelligent machines, and I would chuckle knowingly and give them a statistics problem," he said. "But I now believe intelligence is something we can re-create in our lifetime."

In the days that followed his Japanese lunch with Larry Page, as he typed up a formal pitch for the Google founder, this became a pillar of his proposal. He told Page that deep learning would not only provide image recognition and machine translation and natural language understanding, but would also push machines toward true intelligence. Before the year was out, the project was approved. It was called Project Marvin, in a nod to Marvin Minsky. Any irony was unintended.

GOOGLE was headquartered in Mountain View, California, about forty miles south of San Francisco along Highway 101, at the southernmost edge of San Francisco Bay. The main campus sat atop a hill just off the highway. There, a group of red-blue-and-yellow-themed buildings encircled a grassy courtyard that included a sand-filled volleyball court and a metal dinosaur statue. When Andrew Ng joined Google in early 2011, this was not where he worked. He worked inside Google X, which had set up shop in a building elsewhere in Mountain View, at the fringes of the company's sprawling Northern California operation. But soon after joining the company, he and Thrun made a trip to the campus on the hill so that they could meet with the head of Google Search. Looking for the budget, resources, and political capital needed to explore Ng's ideas, Thrun set up meetings with several of the leading figures inside Google, and the first was with Amit Singhal, who had overseen the Google search engine for nearly a decade. Ng gave him the same pitch he gave Larry Page, except he focused squarely on the search engine, the jewel in the company's crown. As successful as it had been over the years, becoming the world's primary gateway to the Internet, Google Search answered queries in a simple way: It responded to keywords. If you searched on five words and then scrambled them and searched again, you would probably get the same results each time. But Ng told Singhal that deep learning could improve his search engine in ways that would never be possible without it. By analyzing millions of Google searches, looking for patterns in what people clicked on and what they didn't, a neural network could learn to deliver something much closer to what people were actually looking for. "People could ask real questions, not just type in keywords," Ng said.

Singhal wasn't interested. "People don't want to ask questions. They want to type in keywords," he said. "If I tell them to ask questions, they'll just be confused." Even if he had wanted to move beyond keywords, he was fundamentally opposed to the idea of a system that learned behavior on such a large scale. A neural network was a "black box." When it made a decision, like choosing a search result, there was no way of knowing exactly *why it made that decision*. Each decision was based on days or even weeks of calculations that ran across dozens of computer chips. No human being could ever wrap their head around everything a neural network had learned. And changing what it had learned was far from trivial, requiring new data and a whole new round of trial and error. After a decade spent running Google Search, Singhal didn't want to lose control over the way his search engine operated. When he and his fellow engineers made changes to their search engine, they knew exactly what they were changing, and they could explain these changes to anyone who asked. That would not be the case with a neural network. Singhal's message was unequivocal. "I don't want to talk to you," he said.

Ng also met with the heads of Google's image search and video search services, and they turned him down, too. He didn't really find a collaborator until he and Jeff Dean walked into the same microkitchen, the very Googley term for the communal spaces spread across its campus where its employees could find snacks, drinks, utensils, microwave ovens, and maybe even a little conversation. Dean was a Google legend.

The son of a tropical disease researcher and a medical anthropologist, Jeff Dean grew up across the globe. His parents' work took the family from Hawaii, where he was born, to Somalia, where he helped run a refugee camp during his middle school years. As a high school senior in Atlanta, Georgia, where his father worked for the Centers for Disease Control and Prevention, he built a software tool for the CDC that helped researchers collect disease data, and nearly four decades later, it remained a staple of epidemiology across the developing world. After graduate school, where he studied computer science at its foundational levels—the "compilers" that turn software code into something the computer can understand—he joined a Silicon Valley research lab run by the Digital Equipment Corporation, and as the influence of this onetime giant of the computer industry waned, he was among the top DEC researchers who streamed into Google just as the company was taking off. Google's early success is often attributed to PageRank, the search algorithm developed by Larry Page while he and his cofounder, Sergey Brin, were graduate students at Stanford. But the slim, square-jawed, classically handsome Dean, who spoke with a polite shyness and a slight lisp, was just as important to the company's rapid rise—if not more so. He and a handful of other engineers built the sweeping software systems that underpinned the Google search engine, systems that ran across thousands of computer servers and multiple data centers, allowing PageRank to instantly serve millions of people with each passing second. "His expertise was in building a system out of millions of computers that behaved like one," says Sebastian Thrun. "No one in the history of computing had ever done that."

Among engineers, Dean was revered like few others in Silicon Valley. "It was the topic of lunch conversation when I was a young engineer. We would sit around and talk about the *mightiness*," remembers Kevin Scott, an early Googler who would go on to become the chief technology officer at Microsoft. "He had this uncanny ability to take these very, very complicated bits and pieces of technology and devolve them down to their essence." One April Fools' Day—a sacred occasion in the early years of Google—a website appeared on the company's private network offering a list of "Jeff Dean Facts," a riff on the "Chuck Norris Facts" that bounced around the Internet in ironic praise of the '80s action movie star:

> Jeff Dean once failed a Turing test when he correctly identified the 203rd Fibonacci number in less than a second.
>
> Jeff Dean compiles and runs his code before submitting, but only to check for compiler and CPU bugs.
>
> Jeff Dean's PIN is the last 4 digits of pi.
>
> The speed of light in a vacuum used to be about 35 mph. Then Jeff Dean spent a weekend optimizing physics.

Other Googlers were encouraged to add their own Facts, and many did. Kenton Varda, the young engineer who created the site, was careful to hide his identity, but after piecing together a few digital clues buried in the Google server logs, Dean tracked him down and sent a note of thanks. What began as an April Fools' joke grew into Google mythology, a tale often repeated both inside the company and out.

Andrew Ng knew that Jeff Dean would bring his project a level of technical expertise that few others could, as well as the political capital that would help the project thrive inside the company. So their meeting in the microkitchen—when Dean asked what Ng was doing at Google and Ng, in his whispery voice, said he was building neural networks—was a crucial one. According to company lore, it was a moment of happenstance that sparked the creation of the Google AI lab. But Dean had actually emailed Ng before the meeting. And from his earliest days at the company, Ng knew his project depended on the interest of Jeff Dean. His constant concern was getting Dean on board—and keeping him there. What he didn't know was that Dean had a history with neural networks. Almost ten years older than Ng, Dean had explored the idea as an undergraduate at the University of Minnesota in the early '90s, during the first renaissance in connectionist research. For his senior thesis, he trained a neural network on a sixty-four-processor machine called "caesar" that seemed enormously powerful at the time but was nowhere close to what the technology would ultimately need to do something useful. "I felt that by parallelizing computations across sixty-four processors, maybe we'd be able to do interesting things," he says. "But I was a bit naïve." He needed a million times more computing power, not sixty. So when Ng said he was working on neural networks, Dean knew exactly what that meant. In fact, two other Googlers, including a neuroscientist named Greg Corrado, were already exploring the idea. "We have a lot of computers at Google," he told Ng in his typically straightforward way. "Why don't we train really big neural networks?" After all, this was Dean's expertise—pooling computing power from hundreds and even thousands of machines and applying them to a single problem. That winter, he set up an extra desk inside Google X and embraced Ng's project during his "20 percent time"—the one day a week Googlers traditionally spend on side projects. In the beginning, Project Marvin was just another experiment, with Ng, Dean, and Corrado lending only some of their attention to the effort.

They built a system that echoed a very human pastime of the early 2010s: It looked at cats in YouTube videos. Drawing on the power of over sixteen thousand computer chips spread across a Google data center, it analyzed millions of these videos and taught itself to recognize a cat. The results weren't nearly as accurate as those of the leading image recognition tools of the day, but they were a step forward in the sixty-year evolution of neural networks. Ng, Dean, and Corrado published their research the following summer in what, among AI specialists, became known as the Cat Paper. The project also appeared in the pages of the *New York Times*, where it was described as a "simulation of the human brain." This was how the researchers viewed their work. Dean and Corrado, the neuroscientist, eventually committed all their time to Ng's project. They also hired additional researchers from Stanford and Toronto as the project graduated from Google X to a dedicated AI lab, Google Brain.

The rest of the industry, and even parts of Google Brain, didn't quite realize what was about to happen. Just as the lab reached this key moment, Andrew Ng decided to leave. He had another project in the works, and it needed his attention. He was building a start-up, Coursera, that specialized in MOOCs, or Massive Open Online Courses, a way of delivering a university education over the Internet. In 2012, it was one of those Silicon Valley ideas that entrepreneurs, investors, and journalists were convinced would completely change the world. At the same moment, Sebastian Thrun was creating a similar start-up called Udacity. Neither project, though, could begin to compete with what was about to unfold inside Google Brain.

In a roundabout way, Ng's departure catalyzed the project. Before he left, he recommended a replacement: Geoff Hinton. Years later, with the benefit of hindsight, it seemed like the natural step for everyone involved. Hinton was not only a mentor to Ng, he had seeded the lab's first big success when he sent Navdeep Jaitly to Google a year earlier, a success that realized a technology Hinton had nurtured for decades. But when Google approached him in the spring of 2012, he had no interest in leaving the University of Toronto. He was a sixty-four-year-old tenured professor overseeing a long line of graduate students and postdocs. So he merely agreed to spend the summer at the new lab. Owing to the idiosyncrasies of Google's employment rules, the company brought him in as an intern, alongside dozens of college students on summer break. He felt like an oddity during orientation week, when he seemed to be the only one who didn't know that an LDAP was a way of logging in to the Google computer network. "After not very many minutes, they decided to take one of the four instructors and just have him stand next to me," he remembers. But during this orientation, he also noticed another group of people who were slightly out of place: several executive types and their personal assistants, who all seemed to be smiling from ear to ear. At lunch one day, Hinton approached them to ask why they were part of the orientation, and they said their company had just been acquired by Google. Selling a company to Google, he thought, was a good way to get a very big grin on your face.

By that summer, Google Brain had expanded into a team of more than ten researchers after moving into a building across the courtyard from the one that housed Larry Page and the rest of the executive team. Hinton knew one of the researchers, a former Toronto postdoc named Marc'Aurelio Ranzato, and he was impressed with Jeff Dean. He compared Dean to Barnes Wallis, the twentieth-century scientist and inventor portrayed in the classic British war movie *The Dam Busters*. At one point in the film, Wallis asks a government official for a Wellington bomber plane. He needs a way of testing a bomb that bounces across the water, a seemingly ridiculous idea that no one thinks will work. The official resists, explaining that there's a war on. Wellington bombers are hard to come by. "They're worth their weight in gold," the official says. But when Wallis points out that he was the one who designed the Wellington bomber, the official gives him what he wants. During Hinton's summer internship, one project ran into a cap Google had placed on available computing power. So the researchers told Jeff Dean, and he ordered up another $2 million worth. He had built the Google infrastructure, and that meant he could use it as he saw fit. "He created a kind of canopy where the Brain team could operate, and we didn't have to worry about anything else," Hinton says. "If you needed something, you asked Jeff and he got it." What was odd about Dean, Hinton thought, was that unlike most people so intelligent and so powerful, he wasn't driven by ego. He was always willing to collaborate. Hinton compared him to Isaac Newton, except that Newton was an asshole: "Most smart people—someone like Newton, for example—hold grudges. Jeff Dean doesn't seem to have that element to his personality."

The irony was that the lab's approach was all wrong. It was using the wrong kind of computing power—and the wrong kind of neural network. Navdeep Jaitly's speech system had been successfully trained on GPU chips. Dean and the other founders of Google Brain, however, trained their systems across the machines that underpinned Google's global network of data centers, and these used thousands of central processing units, or CPUs (the chips at the heart of every computer), rather than GPUs. At one point, Sebastian Thrun had lobbied the company's head of infrastructure to install GPU machines inside its data centers, but he was turned down on the grounds that this would complicate the company's data center operation and drive up costs. When Jeff Dean and his team unveiled their methods at one of the big AI conferences, Ian Goodfellow, still a student at the University of Montreal, stood up from his seat in the audience and chided them for not using GPUs—though he would soon regret that he had so blithely and so publicly criticized Jeff Dean. "I had no idea who he was," Goodfellow says. "Now I kind of worship him."

The system, known as DistBelief, was also running the wrong kind of neural network. Typically, researchers had to label each image before it could help train a neural network. They had to identify each cat as a cat, drawing a digital "bounding box" around each individual animal. But Google's Cat Paper detailed a system that could learn to recognize cats and other objects from raw *unlabeled* images. Although Dean and his collaborators showed they could train a system without labeling images, it turned out that neural networks were far more accurate, reliable, and efficient if the data they were given was labeled. That fall, when Hinton returned to the University of Toronto after his brief internship at Google, he and two of his students demonstrated, quite clearly, that Google was on the wrong path. They built a system that analyzed labeled images and learned to recognize objects with an accuracy well beyond any technology anyone had ever previously built, showing that machines were more efficient when humans pointed them in the right direction. Neural networks learned in more powerful ways if someone showed them exactly where the cats were.

IN the spring of 2012, Geoff Hinton phoned Jitendra Malik, the University of California–Berkeley professor who had publicly attacked Andrew Ng over his claims that deep learning was the future of computer vision. Despite the success of deep learning with speech recognition, Malik and his colleagues questioned whether the technology would ever master the art of identifying images.

And because he was someone who generally assumed that incoming calls were arriving from telemarketers trying to sell him something, it was surprising that he even picked up the phone. When he did, Hinton said: "I hear you don't like deep learning." Malik said this was true, and when Hinton asked why, Malik said there was no scientific evidence to back any claim that deep learning could outperform any other technology on computer vision. Hinton pointed to recent papers showing that deep learning worked well when identifying objects on multiple benchmark tests. Malik said these datasets were too old. No one cared about them. "This is not going to convince anyone who doesn't share your ideological predilections," he said. So Hinton asked what *would* convince him.

At first, Malik said deep learning would have to master a European dataset called PASCAL. "PASCAL is too small," Hinton told him. "To make this work, we need a lot of training data. What about ImageNet?" Malik told him they had a deal. ImageNet was an annual contest run by a lab at Stanford, about forty miles south of Berkeley. The lab had compiled a vast database of carefully labeled photos, everything from dogs to flowers to cars, and each year, researchers across the globe competed to build a system that could recognize the most images. Excelling at ImageNet, Hinton thought, would certainly clinch the argument. What he didn't tell Malik was that his lab was already building a neural network for the upcoming competition, and that thanks to two of his students—Ilya Sutskever and Alex Krizhevsky—it was nearly finished.

Sutskever and Krizhevsky were typical of the international nature of AI research. Both were born in the Soviet Union before moving to Israel and then Toronto. But otherwise, they were very different. Ambitious, impatient, and even pushy, Sutskever had knocked on Hinton's office door nine years earlier while still an undergraduate at the University of Toronto trying to pay the bills with a job making french fries at a local fast-food restaurant. When the door opened, he immediately asked, in his clipped Eastern European accent, if he could join Hinton's deep learning lab.

"Why don't you make an appointment and we could talk about it?" Hinton said.

"Okay," Sutskever said. "How about now?"

So Hinton invited him in. Sutskever was a mathematics student, and in those few minutes, he seemed like a sharp one. Hinton gave him a copy of the backpropagation paper—the paper that had finally revealed the potential of deep neural networks twenty-five years earlier—and told him to come back once he'd read it. Sutskever returned a few days later.

"I don't understand it," he said.

"It's just basic calculus," Hinton said, surprised and disappointed.

"Oh, no. What I don't understand is why you don't take the derivatives and give them to a sensible function optimizer."

"It took me five years to think of that," Hinton told himself. So he handed the twenty-one-year-old a second paper. Sutskever came back a week later.

"I don't understand it," he said.

"Why not?"

"You train a neural network to solve a problem, and then, if you want to solve a different problem, you start again with another neural network and you train it to solve a different problem. You ought to have one neural network that trains on all these problems."

Realizing that Sutskever had a way of reaching conclusions that even seasoned researchers took years to find, Hinton invited him to join his lab. His education was well behind the rest of the students when he first joined—maybe years behind, Hinton thought—but he caught up in a matter of weeks. Hinton came to see him as the only student he ever taught who had more good ideas than he did, and Sutskever—who kept his dark hair closely cropped and always seemed to be scowling, even when he wasn't—fed these ideas with an almost manic energy. When the big ideas came, he would punctuate the moment with handstand push-ups in the middle of the Toronto apartment he shared with George Dahl. "Success is guaranteed," he would say. In 2010, after reading a paper published by Jürgen Schmidhuber's lab in Switzerland, he stood in a hallway with several other researchers and announced that neural networks would solve computer vision, insisting it was merely a matter of someone doing the work.

Hinton and Sutskever—the idea men—saw how neural networks could crack ImageNet, but they needed Alex Krizhevsky's skills to get there. Laconic and retiring, Krizhevsky wasn't one for the big idea, but he was an unusually talented software engineer with a knack for building neural networks. Leaning on experience, intuition, and a little luck, researchers like Krizhevsky built these systems through trial and error, working to coax a result from hours or even days of computer calculations they could never perform on their own. They assigned tiny mathematical operations to dozens of digital "neurons," fed thousands of dog photos into this artificial neural network, and hoped that, after the many hours of calculation, it would learn to recognize a dog. When it didn't, they adjusted the math and tried again—and again—until it worked. Krizhevsky was a master of what some called a "dark art." But more important, at least in the moment, he had a way of squeezing every last bit of speed from a machine filled with GPU chips, which were still an unusual breed of computer hardware. "He is very good at neural net research," Hinton says. "But he's an amazing software engineer."

Krizhevsky had not even heard of ImageNet before Sutskever

mentioned it, and once he knew what the plan was, he wasn't nearly as enthusiastic as his labmate about the possibilities. Sutskever spent weeks massaging the data so that it would be particularly easy to work with, while Hinton told Krizhevsky that every time he improved the performance of their neural network by 1 percent, he could have an extra week to write his "depth paper," a sweeping school project that was already weeks late. ("That was a joke," Krizhevsky says. "He may have thought it was a joke. But it wasn't," Hinton says.)

Krizhevsky, who still lived with his parents, trained his neural network on a computer in his bedroom. As the weeks passed, he coaxed more and more performance from the machine's two GPU cards, and that meant he could feed his neural network with more and more data. The University of Toronto, Hinton liked to say, didn't even have to pay for the electricity. Each week, Krizhevsky would start the training, and with each passing hour, on the computer screen in his bedroom, he could watch its progress—a black screen filled with white numbers counting upward. At the end of the week, he would test the system on a new set of images. It would fall short of the goal, so then he would hone the GPU code and adjust the weights of the neurons and train for another week. And another. And another. Each week, too, Hinton would oversee a gathering of the students in his lab. It operated like a Quaker meeting. People would just sit there until someone decided to speak up and share what they were working on and what progress they were seeing. Krizhevsky rarely spoke up. But when Hinton coaxed the results out of him, a real sense of excitement would build in the room. "Every week, he'd try to get Alex Krizhevsky to say a little bit more," remembers Alex Graves, another member of the lab in those years. "He knew how big this was." By the fall, Krizhevsky's neural network had passed the current state of the art. It was nearly twice as accurate as the next best system on Earth. And it won ImageNet.

Krizhevsky, Sutskever, and Hinton went on to publish a paper describing their system (later christened AlexNet), which Krizhevsky unveiled at a computer vision conference in Florence, Italy, near the end of October. Addressing an audience of more than a hundred researchers, he described the project in his typically soft, almost apologetic tones. Then, as he finished, the room erupted with argument. Rising from his seat near the front of the room, a Berkeley professor named Alexei Efros told the room that ImageNet was not a reliable test of computer vision. "It is not like the real world," he said. It might include hundreds of photos of T-shirts and AlexNet might have learned to identify these T-shirts, he told the room, but the T-shirts were neatly laid out on tables without a wrinkle, not worn by real people. "Maybe you can detect these T-shirts in an Amazon catalog, but that will not help you in detecting real T-shirts." His Berkeley colleague Jitendra Malik, who had told Hinton he would change his mind about deep learning if a neural network won ImageNet, said he was impressed but would withhold judgment until the technology was applied to other datasets. Krizhevsky didn't have the chance to defend his work. That role was taken by Yann LeCun, who stood up to say this was an unequivocal turning point in the history of computer vision. "This is proof," he said, in a booming voice from the far side of the room.

He was right. And after facing years of skepticism over the future of neural networks, he was vindicated. In winning ImageNet, Hinton and his students had used a modified version of LeCun's creation from the late 1980s: the convolutional neural network. But for some students in LeCun's lab, it was also a disappointment. After Hinton and his students published the AlexNet paper, LeCun's students felt a deep sense of regret descend on their own lab—a sense that after thirty years of struggle, they had stumbled at the last hurdle. "Toronto students are faster than NYU students," LeCun told Efros and Malik as they discussed the paper later that night.

In the years that followed, Hinton compared deep learning to the theory of continental drift. Alfred Wegener first proposed the idea in 1912, and for decades it was dismissed by the geological community, partly because Wegener was not a geologist. "He had evidence, but he was a climatologist. He was not 'one of us.' He was laughed at," Hinton says. "It was the same with neural nets." There was plenty of evidence that neural networks could succeed across a wide variety of tasks, but it was ignored. "It was just asking too much to believe that if you started with random weights and you had lots of data and you followed the gradient, you would create all these wonderful representations. Forget it. Wishful thinking."

In the end, Alfred Wegener was vindicated, but he did not live to enjoy the moment. He died on an expedition to Greenland. With deep learning, the pioneer who did not live to see the moment was David Rumelhart. In the '90s, he developed a degenerative brain condition called Pick's disease that began to destroy his judgment. Before he was diagnosed, he divorced his wife after a long and happy marriage and quit his job for a lesser one. He eventually moved to Michigan, where he was cared for by his brother, and he died in 2011, a year before AlexNet. "If he had lived," Hinton says, "he would have been the major figure."

The AlexNet paper would become one of the most influential papers in the history of computer science, with over sixty thousand citations from other scientists. Hinton liked to say it was cited fifty-nine thousand more times than any paper his father ever wrote. "But who's counting?" he would ask. AlexNet was a turning point not just for deep learning but for the worldwide technology industry. It showed that neural networks could succeed in multiple areas—not just speech recognition—and that GPUs were essential to this success. It shifted both the software and hardware markets. Baidu recognized the importance, after the deep learning researcher Kai Yu explained the moment to the CEO, Robin Li. So did Microsoft, after Li Deng won the support of an executive vice president named Qi Lu. And so did Google.

It was at this pivotal moment that Hinton created DNNresearch, the company they would auction off in a Lake Tahoe hotel room that December for $44 million. When it came time to split the proceeds, the plan had always been to divide the money equally among the three of them. But at one point, the two graduate students told Hinton he deserved a larger share: 40 percent. "You're giving away an awful lot of money," he told them. "Go back to your rooms and sleep on it."

When they returned the next morning, they insisted he take the larger share. "It tells you what kind of people they are," Hinton says. "It doesn't tell you what kind of person I am."

6

AMBITION

"LET'S REALLY GO BIG."

For Alan Eustace, acquiring DNNresearch was just the beginning. As Google's head of engineering, he was intent on cornering the worldwide market for deep learning researchers—or at least coming close. Larry Page, the CEO, had made this a priority several months earlier, when he and the rest of the Google executive team had gathered for a strategy meeting on an (undisclosed) island in the South Pacific. Deep learning was about to change the industry, Page told his lieutenants, and Google needed to get there first. "Let's really go big," he said. Eustace was the only one in the room who really knew what he was talking about. "They all stepped back," Eustace remembers. "I didn't." In that moment, Page gave Eustace free rein to secure any and all of the leading researchers in what was still a tiny field, potentially hundreds of new hires. He had already landed Hinton, Sutskever, and Krizhevsky from the

University of Toronto. Now, in the last days of December 2013, he was flying to London in pursuit of DeepMind.

Founded around the same time as Google Brain, DeepMind was a start-up dedicated to an outrageously lofty goal. It aimed to build what it called "artificial general intelligence"—AGI—technology that could do anything the human brain could do, only better. That endgame was still years, decades, or perhaps even centuries away, but the founders of this tiny company were confident it would one day be achieved, and like Andrew Ng and other optimistic researchers, they believed that many of the ideas brewing at labs like the one at the University of Toronto were a strong starting point. Though it lacked the deep pockets of its main rivals, DeepMind had joined the bidding for Hinton's company, and it had amassed what may have been the most impressive collection of young AI researchers in the world, even compared to the rapidly growing roster at Google. As a result, this would-be poacher had become a target for other poachers, including Google's biggest rivals: Facebook and Microsoft. This gave Eustace's trip an added sense of urgency. Eustace, Jeff Dean, and two other Googlers planned to spend two days at the DeepMind offices near Russell Square in central London so they could vet the lab's technology and talent, and they knew that one more Googler should join them: Geoff Hinton. But when Eustace asked Hinton to join their transatlantic scouting trip, he politely declined, saying his back wouldn't allow it. The airlines made him sit down during takeoff and landing, he said, and he no longer sat down. At first, Eustace took Hinton's refusal at face value. Then he said he would find a solution.

Eustace was not just an engineer. A trim, straight-backed man who wore rimless glasses, he was also a pilot, skydiver, and all-around thrill seeker who choreographed each new thrill with the same cold rationality he applied to building a computer chip. He would soon set a world record when he donned a pressure suit and jumped from a balloon floating in the stratosphere, twenty-five miles above the Earth. Just recently, he and several other skydivers had parachuted from a Gulfstream jet—something no one else had ever done—and this gave him an idea. Before any of them could make the leap, someone had to open the door at the back of the plane, and just to make sure they didn't go tumbling into the great beyond before they were ready for the jump, they wore full-body mountain-climbing harnesses rigged with two long black straps that hooked into metal rings on the inside walls of the plane. If Google rented a private jet, Eustace decided, they could put a harness on Hinton, lay him on a bed secured to the floor, and hook him into the jet in much the same way. And that's what they did. They took off for London in a private Gulfstream with Hinton lying on two seats folded flat into a makeshift bed, two taught straps holding him in place. "Everyone was pleased with me," Hinton says. "It meant they could fly on the private jet, too."

Based in San Jose, California, this private plane was often rented by Google and other Silicon Valley tech giants, and with each company, the flight crew would change the lighting scheme inside the cabin to match the company's corporate logo. The lights were blue, red, and yellow when the Googlers boarded on a Sunday in December 2013. Hinton wasn't sure what the harness meant for his personal safety, but he felt it at least kept him from tumbling across the plane and crashing headlong into his Google colleagues during takeoff and landing. They landed in London that evening, and the following morning, Hinton walked into DeepMind.

DEEPMIND was led by an eclectic array of powerful minds. Two of them, Demis Hassabis and David Silver, had met as undergraduates at Cambridge, but they originally crossed paths at a youth chess tournament near Silver's hometown on the east coast of England. "I knew Demis before he knew me," Silver says. "I would see him turn up in my town, win the competition, and leave." The son of a Chinese-Singaporean mother and a Greek-Cypriot father who ran a North London toy store, Hassabis was at one point the second-highest-rated under-fourteen chess player in the world. But his talents weren't limited to chess. He graduated from Cambridge with a First in computer science, and he had a way of mastering most games of the mind. In 1998, aged twenty-one, he entered the Pentamind competition, held at London's Royal Festival Hall, in which players from around the world competed across five games of their choice. The competition included everything from chess and Go to Scrabble, backgammon, and poker—and Hassabis won it outright. He went on to win four more times over the next five years. The one year he didn't win, he didn't enter. "Despite its rarefied image, Mind Sports is as competitive as any other sport," he said in an online diary after winning the competition a second time. "At the highest level, everything goes. Sledging, shaking of tables and all manner of skullduggery is par for the course. The junior Chess tournaments I used to attend had wooden boards under the tables preventing competitors from kicking each other. Don't be fooled—this is warfare." Geoff Hinton later said that Hassabis could lay claim to being the greatest games player of all time, before adding, pointedly, that his prowess demonstrated not only his intellectual heft but also his extreme and unwavering desire to win. After his success in the Pentamind, Hassabis won the world team championship in Diplomacy, a board game set in Europe before the First World War in which the top players rely on the analytical and strategic skills of a chess player while also drawing on the guile needed to negotiate, cajole, and connive their way to victory. "He's got three things," Hinton says. "He's very bright, he's very competitive, and he's very good at social interactions. That's a dangerous combination."

Hassabis had two obsessions. One was designing computer games. In his gap year, he helped celebrated British designer Peter Molyneux create Theme Park, in which players build and operate a sprawling digital simulation of a Ferris-wheel-and-rollercoaster amusement park. It sold an estimated 10 million copies, helping to inspire a whole new breed of game—"sims" that re-created huge swaths of the physical world. His other obsession was AI. One day, he believed, he would create a machine that could mimic the brain. In the years to come, as he built DeepMind, these two obsessions would merge in a way few others expected.

In fellow Cambridge undergraduate David Silver he had found a kindred spirit. After university, the two men launched a computer games company called Elixir. As Hassabis built this London start-up, he kept a running online diary of life both inside the company and out (mostly in). This was a promotional vehicle ghost-written by one of his designers—a way of generating interest in his company and its games—but he was baldly honest in places, revealing his geek charm, his cunning, and his steely determination to win. At one point, he described a meeting with Eidos, the big-name British publisher that agreed to distribute his company's first game. It was vitally important, Hassabis said, for a game developer to establish a deep sense of trust with its publisher, and he felt he succeeded during this long meeting inside his offices in London. But as the meeting ended, the Eidos chairman—Ian Livingstone, a man who would later be named a Commander of the Order of the British Empire for his services to the industry—noticed a foosball table across the room and challenged Hassabis to a game. Hassabis stopped to think about whether he should lose the contest, just to keep his publisher happy, before deciding he had no choice but to win. "Ian's no mean player—he is rumored to have been Hull University doubles champion with Steve Jackson—but what a terrible situation to put me in," Hassabis said. "A battling defeat at the hands of Eidos's Chairman (in the face of the superior table-footballing skills) would've been just the ticket. You've got to draw the line somewhere, though. A game is a game, after all. I won 6–3."

The diary seemed to look beyond Elixir, toward his next venture. The first entry began with him sitting in a plush chair at home, listening to the musical score from *Blade Runner* (track twelve, "Tears in Rain," on continuous repeat). Much as Stanley Kubrick inspired a young Yann LeCun in the late '60s, Ridley Scott had captured the imagination of a young Hassabis in the early '80s with this latter-day sci-fi classic, in which a scientist and his imperious corporation build machines that behave like humans. As smaller game developers were squeezed out of the market and Hassabis folded Elixir, he resolved to build another start-up. This one, he decided, would be far more ambitious than the last, returning to his roots in computer science—and science fiction. He resolved, in 2005, to build a company that could re-create human intelligence.

He knew he was years away from even taking the first small step. Before actually launching a company, he started a neuroscience PhD at University College London, hoping to better understand the brain before trying to rebuild it. "My sojourn into academia was always going to be temporary," he says. David Silver also returned to academia, but not as a neuroscientist. He slipped into an adjacent field—artificial intelligence—at the University of Alberta in Canada. The way their research diverged before coming back together with DeepMind was indicative of the relationship between neuroscience and artificial intelligence among at least some of the researchers who drove the big changes in AI during these years. No one could truly understand the brain and no one could re-create it, but some believed these two efforts would ultimately pull each other up by the bootstraps. Hassabis called it "a virtuous circle."

At UCL, he explored the intersection of memory and imagination in the brain. With one paper, he studied people who developed amnesia after brain damage and were unable to remember the past, showing that they also struggled to imagine themselves in new situations, like a trip to the shopping center or a vacation at the beach. Recognizing, storing, and recalling images was somehow related to *creating them.* In 2007, *Science*, one of the world's leading academic journals, named this one of the top ten scientific breakthroughs of the year. But it was just another stepping-stone. After finishing his PhD, Hassabis started a postdoc at the Gatsby Unit, a University College London lab that sat at the intersection of neuroscience and AI. Funded by David Sainsbury, the British supermarket magnate, its founding professor was Geoff Hinton.

Hinton left his new post after only three years, returning to his professorship in Toronto while Hassabis was still running his games company. Several years would pass before they met, and then it was only in passing. Instead, Hassabis made common cause with a fellow Gatsby researcher named Shane Legg. At the time, as he later recalled, AGI wasn't something that serious scientists discussed out loud, even at a place like the Gatsby. "It was basically eye-rolling territory," he says. "If you talked to anybody about general AI, you would be considered at best eccentric, at worst some kind of delusional, nonscientific character." But Legg, a New Zealander who had studied computer science and mathematics while practicing ballet on the side, felt the same way as Hassabis. He dreamed of building *superintelligence*—technology that could eclipse the powers of the brain—though he worried that these machines could one day endanger the future of humanity. Superintelligence, he had said in his thesis, could bring unprecedented wealth and opportunity—or lead to a "nightmare scenario" that threatened the very existence of humankind. Even if there was only a tiny possibility of building superintelligence, he believed, researchers had to consider the consequences. "If one accepts that the impact of truly intelligent machines is likely to be profound, and that there is at least a small probability of this happening in the foreseeable future, it is only prudent to try to prepare for this in advance. If we wait until it seems very likely that intelligent machines will soon appear, it will be too late to thoroughly discuss and contemplate the issues involved," he wrote. "We need to be seriously working on these things now." His larger belief was that the brain itself would provide a map for building superintelligence, and this was what took him to the Gatsby Unit. "That seemed like a very natural place to go," he says—a place where he could explore what he called the "connections between the brain and machine learning."

Years later, when asked to describe Shane Legg, Geoff Hinton compared him to Demis Hassabis: "He's not as bright, not as competitive, and not as good at social interactions. But then, that applies to almost everybody." Even so, in the years to come, Legg's ideas were nearly as influential as his more celebrated partner.

Hassabis and Legg held the same ambition. They wanted to, in their words, "solve intelligence." But they disagreed on the best way to get there. Legg suggested they begin in academia, while Hassabis said they had no choice but to work in industry, insisting this was the only way of generating the resources they would need for such an extreme task. He knew the academic world, and after his time at Elixir, he knew the business world, too. He didn't want to build a start-up for the sake of building a start-up. He wanted to create a company uniquely equipped for the long-term research they hoped to foster. They could raise far more money from venture capitalists than they ever could writing grant proposals as professors, he told Legg, and they could erect the necessary hardware at a speed universities never could. Legg, in the end, agreed. "We didn't actually tell anybody at Gatsby what we were planning," Hassabis says. "They would have thought we were kind of mad."

During their postdoc year, they began spending time with an entrepreneur and social activist named Mustafa Suleyman. When the three of them decided to set up DeepMind, it was Suleyman who provided the financial brains, charged with generating the revenues the company needed to sustain their research. They launched DeepMind in the fall of 2010, its name a nod both to deep learning and to neuroscience—and to the Deep Thought supercomputer that calculated the Ultimate Question of Life in the British sci-fi novel *The Hitchhiker's Guide to the Galaxy.* Hassabis, Legg, and Suleyman would each stamp a unique point of view on a company that looked toward the horizons of artificial intelligence but also aimed to solve problems in the nearer term, while openly raising concerns over the dangers of this technology in both the present and the future. Their stated aim—contained in the first line of their business plan—was artificial general intelligence. But at the same time, they told anyone who would listen, including potential investors, that this research could be dangerous. They said they would never share their technology with the military, and in an echo of Legg's thesis, they warned that superintelligence could become an existential threat.

They approached DeepMind's most important investor even before the company was founded. In recent years, Legg had joined an annual gathering of futurists called the Singularity Summit. "The

Singularity" is the (theoretical) moment when technology improves to the point where it can no longer be controlled by the human race. The founders of this tiny conference belonged to an eclectic group of fringe academics, entrepreneurs, and hangers-on who believed this moment was on the way. They were intent on exploring not only artificial intelligence but life-extension technologies and stem-cell research and other disparate strands of futurism. One of the founders was a self-educated philosopher and self-described artificial intelligence researcher named Eliezer Yudkowsky, who had introduced Legg to the idea of superintelligence in the early 2000s when they were working with a New York–based start-up called Intelligensis. But Hassabis and Legg had their eyes on one of the other conference founders: Peter Thiel.

In the summer of 2010, Hassabis and Legg arranged to address the Singularity Summit, knowing that each speaker would be invited to a private party at Thiel's town house in San Francisco. Thiel had been part of the team that founded PayPal, the online payments service, before securing an even bigger reputation—and an even bigger fortune—as an early investor in Facebook, LinkedIn, and Airbnb. If they could get inside his town house, they felt, they could pitch him their company and lobby for investment dollars. Thiel not only had the money, he had the inclination. He was someone who believed in extreme ideas, even more so than the typical Silicon Valley venture capitalist. After all, he was funding the Singularity Summit. In the coming years, he would, unlike many Valley moguls, put his full weight behind Donald Trump, both before and after the 2016 presidential election. "We needed someone who was crazy enough to fund an AGI company," Legg says. "He was deeply contrarian—still is about everything. Most of the field thought that we shouldn't be doing this, so him being deeply contrarian was probably going to play to our advantage."

On the first day of the conference, inside a hotel in downtown San Francisco, Hassabis gave a speech arguing that the best way to build artificial intelligence was to mimic the way the brain worked. He called this "the biological approach," when engineers designed technologies in the image of the brain, be they neural networks or some other digital creation. "We should be focusing on the algorithmic level of the brain," he said, "extracting the kind of representations and algorithms the brain uses to solve the kinds of problems we want to solve with AGI." This was one of the central pillars that would come to define DeepMind. A day later, with his own speech, Shane Legg described another. He told his audience that artificial intelligence researchers needed definitive ways of tracking their progress. Otherwise, they wouldn't know when they were on the right path. "I want to know where we're going," he said. "We need a concept of what intelligence is, and we need a way to measure it." Hassabis and Legg were not just describing how their new company would operate. Their speeches were, more than anything else, a way of getting to Thiel.

Thiel's town house sat on Baker Street, looking out onto the same geese-filled lagoon as the Palace of Fine Arts, a stone castle of a building erected for an art exhibition nearly a hundred years earlier. As Hassabis and Legg walked through the front door and into the living room, they were greeted by a chessboard. Each piece was in place, white arrayed against black, inviting anyone to play. First, they found Yudkowsky, who introduced them to Thiel. But they didn't pitch their company—at least not right away. Hassabis started talking chess. He told Thiel that he, too, was a chess player, and they discussed the enduring power of this ancient game. It had survived for so many centuries, Hassabis said, because of the deep tension between the knight and the bishop, the constant push and pull of their skills and weaknesses. Thiel was charmed enough to invite the two men back the next day so they could pitch their company.

When they returned the following morning, Thiel was dressed in shorts and a T-shirt, dripping with sweat after his daily workout. A butler brought him a Diet Coke before they sat down at the dining room table. Hassabis did the talking, explaining that he was not just a gamer but a neuroscientist, that they were building AGI in the image of the human brain, that they would begin this long quest with systems that learned to play games, and that the ongoing exponential growth in global computing power would drive their technology to far greater heights. It was a pitch that surprised even Peter Thiel. "This might be a bit much," he said. But they kept talking, and the conversation continued over the next several weeks, with both Thiel and the partners at his venture capital firm, the Founders Fund. In the end, his main objection was not that the company was overly ambitious but that it was based in London. It would be harder to keep an eye on his investment, a typical concern for Silicon Valley venture capitalists. Nevertheless, he invested £1.4 million of the initial £2 million in seed funding that gave rise to DeepMind. In the coming months and years, other big-name investors joined in, including Elon Musk, the Silicon Valley kingpin who helped build PayPal alongside Thiel before building the rocket company SpaceX and the electric-car company Tesla. "There's a certain community," Legg says. "He was one of the billionaires who decided to put some money in."

DeepMind snowballed from there. Hassabis and Legg enlisted both Hinton and LeCun as technical advisors, and the start-up quickly hired many of the field's up-and-coming researchers, including Vlad Mnih, who studied with Hinton in Toronto; a Turkish-born researcher, Koray Kavukcuoglu, who worked under LeCun in New York; and Alex Graves, who studied under Jürgen Schmidhuber in Switzerland before a postdoc with Hinton. As they'd told Peter Thiel, the starting point was games. Games had been a proving ground for AI since the '50s, when computer scientists built the first automated chess players. In 1990, researchers marked a turning point when they built a machine called Chinook that beat the world's best checkers player. Seven years later, IBM's Deep Blue supercomputer topped chess grandmaster Garry Kasparov. And in 2011, another IBM machine, Watson, eclipsed the all-time winners on *Jeopardy!* Now a team of DeepMind researchers, led by Mnih, began building a system that could play old Atari games, including '80s classics like Space Invaders, Pong, and Breakout. Hassabis and Legg were adamant that AI should be developed in a way that closely measured its progress, in part because this would help keep tabs on the dangers. Games provided this measurement. Scores were absolute. Results were definitive. "It's the way we plant our flags," Hassabis says. "Where should we go next? Where's the next Everest?" Plus, AI that played games made for a very good demo. Demos sold software—and sometimes companies. This was apparent, even undeniable, by early 2013.

In Breakout, players use a tiny paddle to bounce a ball off a wall of colored bricks. When you hit a brick, it disappears, and you win a few points. But if the ball bounces past your paddle too many times, the game is over. At DeepMind, Mnih and his fellow researchers built a deep neural network that learned the nuances of Breakout through repeated trial and error, playing hundreds of games while keeping close track of what worked and what didn't—a technique called "reinforcement learning." This neural network could master the game in little more than two hours. Within the first thirty minutes, it learned the basic concepts—move toward the ball and hit the ball toward the bricks—though it hadn't yet mastered them. After an hour, it was adept enough to return the ball every time and score points with each hit. And after two hours, it learned a trick that broke the game open. It would hit the ball *behind* the wall of colored bricks, slipping it into a pocket of space where it could bounce around almost endlessly, knocking down brick after brick and scoring point after point without ever returning to the paddle. In the end, the system played with a speed and precision beyond any human.

Shortly after Mnih and his team built this system, DeepMind sent a video to the company's investors at the Founders Fund, including a man named Luke Nosek. Alongside Peter Thiel and Elon Musk, Nosek had originally risen to prominence as part of the team that created PayPal—the so-called "PayPal Mafia." Soon after receiving the video of DeepMind's Atari-playing AI, as Nosek later told a colleague, he was on a private plane with Musk, and as they watched the video and discussed DeepMind, they were overheard by another Silicon Valley billionaire who happened to be on the flight: Larry Page. This was how Page learned about DeepMind, sparking a courtship that would culminate in the Gulfstream flight to London. Page wanted to buy the start-up, even at this early stage. Hassabis wasn't so sure. He had always intended to build his own company. Or, at least, this was what he told his employees. He said that DeepMind would remain independent for the next twenty years, if not more.

THE elevator Hinton and the other Googlers took to the lobby of the DeepMind offices got stuck between floors. As they waited, Hinton worried the delay was preying on those inside the DeepMind offices, many of whom he knew. "This must be embarrassing," he thought. When the elevator finally restarted and the Googlers arrived on the top floor, they were greeted by Hassabis, who led them into a room with a long conference table. He was not so much embarrassed as he was nervous, wary of exposing the lab's research to a company whose outsized resources could accelerate this research in ways the lab never could on its own. He didn't want to expose the company's secrets unless he was sure that he wanted to sell—and that Google wanted to buy. After the Googlers filed into the room, he gave a speech explaining DeepMind's mission. Then several DeepMind researchers revealed at least part of what the lab was exploring, from the concrete to the theoretical. The money moment came from Vlad Mnih, and as usual, it was Breakout.

As Mnih presented the project, an exhausted Geoff Hinton lay on the floor, beside the table where everyone else was sitting. Occasionally, Mnih would see Hinton's hand pop up when he wanted to ask a question. It was, Mnih thought, just like their days in Toronto. As the demo ended, Jeff Dean asked if the system was truly *learning* its Breakout skills. Mnih said it was. It was homing in on particular strategies because they won the most reward—in this case, the most points. This technique—reinforcement learning—was not something Google was exploring, but it was a primary area of research inside DeepMind. Shane Legg had embraced the concept after his postdoctoral advisor published a paper arguing that the brain worked in much the same way, and the company had hired a long list of researchers who specialized in the idea, including David Silver. Reinforcement learning, Alan Eustace believed, allowed DeepMind to build a system that was the first real attempt at general AI. "They had superhuman performance in like half the games, and in some cases, it was astounding," he says. "The machine would develop a strategy that was just a killer."

After the Atari demos, Shane Legg gave a presentation based on his PhD thesis, in which he described a breed of mathematical agent that could learn new tasks in any environment. Vlad Mnih and his team had built agents that could learn new behavior inside games like Breakout and Space Invaders, and what Legg proposed was an extension of this work—beyond games and into more complex digital realms as well as the real world. Just as a software agent could learn to navigate Breakout, a robot could learn to navigate a living room or a car could learn to navigate a neighborhood. Or, in much the same way, one of these agents could learn to navigate the English language. These were far more difficult problems. A game was a contained universe where the rewards were clearly defined. There were points and finish lines. The real world was far more complicated, the rewards harder to define, but this was the path DeepMind laid out for itself. "Shane's thesis," Eustace says, "formed the core of what they were doing."

This was a goal in the distant future, but there would be many small steps along the way, steps that would bring practical applications in the nearer term. As the Googlers looked on, Alex Graves, the son of American parents who grew up in Scotland, demonstrated one of them: a system that wrote "in longhand." By analyzing the patterns that define an object, a neural network could learn to recognize it. If it understood those patterns, it could also *generate* an image of that object. After analyzing a collection of handwritten words, Graves's system could generate *the image of a handwritten word*. The hope was that, by analyzing photos of dogs and cats, this kind of technology could also generate photos of dogs and cats. Researchers called them "generative models," and this, too, was a major area of research at DeepMind.

At a time when the Googles of the world were paying hundreds of thousands of dollars, if not millions, for each researcher, DeepMind was paying the likes of Alex Graves less than $100,000 a year. That was all it could afford. Three years after it was founded, this tiny company still had no revenue. Suleyman and his team were trying to build a mobile app that would use AI to help you sift through the latest couture—fashion editors and writers would occasionally walk through the Russell Square office amid the AI researchers—and a separate group was on the verge of offering a new AI video game in the Apple App Store, but the dollars were not yet coming in. As Graves and other researchers described their work to the visitors from Google, Hassabis knew something had to change.

When the demos ended, Jeff Dean asked Hassabis if he could take a look at the company's computer code. Hassabis balked at first, but then agreed, and Dean sat down at a machine alongside Koray Kavukcuoglu, who oversaw Torch, software the company used to build and train its machine learning models. After about fifteen minutes with the code, Dean knew DeepMind would fit with Google. "It was clearly done by people who knew what they were doing," he says. "I just felt like their culture would be compatible with our culture." By this point, there was little doubt Google would acquire the London lab. Mark Zuckerberg and Facebook had recently joined Google, Microsoft, and Baidu in the race to acquire this kind of talent, and Google was intent on staying ahead. Though Hassabis had long promised his employees that DeepMind would stay independent, he now had no choice but to sell. If DeepMind didn't sell, it would die. "We couldn't really sustain having these $100 billion companies desperate to hire all our top talent," Legg says. "We managed to keep everybody, but this was not sustainable in the long run."

Still, as they negotiated DeepMind's sale to Google, they held on to at least part of the promise Hassabis made to his employees. DeepMind would not stay independent for more than three weeks, much less another twenty years, but he and Legg and Suleyman insisted their Google contract include two conditions that aimed to maintain their ideals. One clause barred Google from using any

DeepMind technology for military purposes. The other said Google was required to create an independent ethics board that would oversee the use of DeepMind's AGI technology, whenever that might arrive. Some who were privy to the contract questioned whether these terms were necessary, and in later years, many across the AI community saw it as a stunt designed to boost DeepMind's sale price. "If they said their technology was dangerous, it seemed more powerful, and they could demand more for it," the voices said. But the founders of DeepMind were adamant the sale would not happen unless these demands were in place, and they continued to fight for the same ideals for years on end.

Before boarding the Gulfstream in California, Hinton had said he would be taking the train back to Canada—a cover story meant to protect the secrecy of the trip to London. On the return flight, the plane made a small detour to Canada to drop him off, landing in Toronto at about the time the train he was supposed to have taken would have arrived. The ruse was maintained. In January, Google announced it was acquiring DeepMind, a fifty-person company, for $650 million. It was another photo finish. Facebook had also bid for the London lab, offering each DeepMind founder twice as much money as they made from the sale to Google.

PART TWO

WHO OWNS INTELLIGENCE?

7

RIVALRY

"HELLO, THIS IS MARK, FROM FACEBOOK."

In late November 2013, Clément Farabet was sitting on the couch in his one-bedroom Brooklyn apartment, typing code into his laptop, when his iPhone rang. The screen read "Menlo Park, CA." When he answered the call, a voice said: "Hello, this is Mark, from Facebook." Farabet was a researcher in the deep learning lab at NYU. A few weeks earlier, he'd been contacted by another Facebook executive, out of the proverbial blue, but he still didn't expect a call from Mark Zuckerberg. In his very direct and unceremonious way, Zuckerberg told Farabet he was traveling to Lake Tahoe for the NIPS conference and asked if they could meet in Nevada for a chat. NIPS was less than a week away and Farabet hadn't planned on making the trip that year, but he agreed to meet Zuckerberg in the penthouse suite at the Harrah's hotel and casino the day before the conference began. When the call ended, he scrambled to arrange a cross-country flight and a place to stay, but he didn't quite realize what was happening until he arrived in Nevada, walked into the Harrah's penthouse, and saw who was sitting on the couch behind the Facebook founder and CEO. It was Yann LeCun.

Zuckerberg wasn't wearing any shoes. Over the next half hour, he paced back and forth across the suite in his stocking feet, calling AI "the next big thing" and "the next step for Facebook." It was a week before the Google entourage flew to London for their meeting with DeepMind, and Facebook was building a deep learning lab of its own. The company had hired LeCun to run the lab days earlier. Now, alongside LeCun and the other man in the room, Facebook chief technology officer Mike "Schrep" Schroepfer, Zuckerberg was recruiting talent for this new venture. Farabet, a Lyon-born academic who specialized in image recognition and had spent years designing computer chips for training neural networks, was just one of the many researchers who walked into the Harrah's penthouse that afternoon to meet with Zuckerberg. "He basically wanted to hire everybody," Farabet says. "He knew the names of every single researcher in the space."

That night, Facebook held a private party in one of the hotel ballrooms. With dozens of engineers, computer scientists, and academics packed into this split-level space, including the balcony overlooking the crowd below, LeCun revealed that the company was opening an AI lab in Manhattan, not far from his office at NYU. "It's a marriage made in heaven—otherwise known as New York City," LeCun said, before raising a glass to "Mark and Schrep." Facebook had already hired a second NYU professor to work alongside LeCun at the new lab—dubbed FAIR, for Facebook Artificial Intelligence Research—and a few more notable names would soon join them, including three researchers poached from Google. But in the end, despite his long history with LeCun, a fellow Frenchman, Clément Farabet did not. He and several other academics were building a deep learning start-up they called Madbits, and he resolved to see it through. Six months later, before this tiny new company had even come close to releasing its first product, it was acquired by Twitter, Silicon Valley's other social networking giant. The battle for talent, already so heated, was only getting hotter.

FACEBOOK'S Silicon Valley headquarters is the corporate campus that feels like Disneyland. Thanks to a rotating team of muralists, sculptors, silk-screeners, and other artists-in-residence, each building, room, hallway, and foyer is carefully decorated with its own colorful extravagance, and in between, the eateries advertise themselves with just as much gusto. Big Tony's Pizza sits on one corner, the Burger Shack on another. Earlier that year, inside Building 16, near Teddy's Nacho Royale, Mark Zuckerberg had sat down with the founders of DeepMind. They shared a notable connection: Peter Thiel, DeepMind's first investor and a Facebook board member. Still, Zuckerberg hadn't been quite sure what to think of the tiny start-up from London. He'd recently met with several other start-ups promising what they called artificial intelligence, and this seemed like one more pitch among many.

Once the meeting had ended, a Facebook engineer named Lubomir Bourdev told Zuckerberg that what they'd heard was more than just hyperbole, that Hassabis and Legg had taken hold of a technology on the rise. "These guys are for real," Bourdev said. A specialist in computer vision, Bourdev was leading a new effort to build a service that could automatically recognize objects in photos and videos posted to Facebook. In the wake of AlexNet, like so many others who'd seen deep learning suddenly eclipse the systems they'd been working on for years, he knew that neural networks would change the way digital technology was built. DeepMind, he told Zuckerberg, was the company Facebook should acquire.

In 2013, that was a strange idea. The wider tech industry, including most engineers and executives at Facebook, hadn't even heard of deep learning and certainly didn't understand its growing importance. More to the point: Facebook was a social networking company. It built Internet technology for the here and now, not "artificial general intelligence" or any other technology unlikely to reach the real world for years to come. The company motto was "Move Fast and Break Things," a slogan repeated almost endlessly on small silk-screened signs spread across the walls of its corporate campus. Facebook ran a social network spanning more than a billion people across the globe, and the company was geared toward expanding and amplifying this service as quickly as possible. It did not do the kind of research DeepMind aimed to do, which was more about exploring new frontiers than moving fast and breaking things. But now, after growing into one of the world's most powerful companies, Zuckerberg was intent on racing the others—Google, Microsoft, Apple, and Amazon—to the Next Big Thing.

This is how the tech industry works. The largest companies are locked in a never-ending race toward the next transformative technology, whatever that might be. Each is intent on getting there first, and if someone beats them to it, then they are under even more pressure to get there, too, without delay. With the acquisition of Geoff Hinton and his start-up, Google had gotten to deep learning first. By the middle of 2013, Zuckerberg decided he had to get there, too, even if he was racing for only second place. It didn't matter that Facebook was merely a social network. It didn't matter that deep learning was not an obvious fit for anything beyond ad targeting and image recognition on this social network. It didn't matter that the company didn't really do long-term research. Zuckerberg was intent on bringing deep learning research to Facebook. That was the job he gave to the man everyone called Schrep.

Mike Schroepfer had joined Facebook five years earlier as its head of engineering, after Zuckerberg's Harvard roommate, company cofounder Dustin Moskovitz, stepped down from the role. He wore black-rimmed glasses and a short, Caesar-like haircut that matched the one worn by Zuckerberg. Almost ten years older than the Facebook CEO, Schrep was a Silicon Valley veteran who'd studied at Stanford alongside a who's who of Silicon Valley veterans. He cut his teeth as the chief technology officer at Mozilla, the company that challenged the monopoly of Microsoft and its Internet Explorer Web browser in the early 2000s. When he moved to Facebook, his primary job was to ensure that the hardware and software supporting the world's largest social network could handle the load as it expanded from a hundred million people to one billion and beyond. But in 2013, when he was promoted to chief technology officer, his priorities changed. Now his task was to push Facebook into entirely new areas of technology, beginning with deep learning. "This is one of many examples where Mark formed a fairly clear view of the future," Schrep later said. What he didn't say was that Google had already come to the same conclusion.

In the end, Zuckerberg and Schroepfer made an unsuccessful bid for DeepMind. Hassabis told colleagues that he felt no chemistry with Zuckerberg, that he didn't quite understand what the Facebook founder was trying to do with DeepMind, and that the lab wouldn't fit with Facebook's growth-obsessed corporate culture. But the bigger issue—for Hassabis, Legg, and Suleyman—was that Zuckerberg didn't share their ethical concerns over the rise of artificial intelligence, in either the near term or the far. He refused to accept a contractual clause that guaranteed DeepMind's technology would be overseen by an independent ethics board.

"We could have made more money—if we had just been going for the money," Legg says. "But we weren't."

Ian Goodfellow, a graduate student at the University of Montreal who would soon become one of the biggest names in the field, was among the many researchers Facebook recruited during this time, and when he met with the Facebook CEO on a visit to company headquarters, he was struck by how much time Zuckerberg spent talking about DeepMind. "I guess it should have occurred to me," Goodfellow says, "that he was thinking of acquiring." But as it eyed the same technological future as Google, Facebook faced a chicken-and-egg problem: It couldn't attract top researchers because it didn't have a research lab, and it didn't have a research lab because it couldn't attract top researchers. The breakthrough was Marc'Aurelio Ranzato. A former professional violinist from Padua, Italy, Ranzato had meandered into the world of technology because he couldn't quite make a living as a musician and thought he could reinvent himself as a recording engineer. Somewhere along the way, this took him into the AI of sound and images. The thin, soft-spoken Italian studied under LeCun at NYU and then Hinton at the University of Toronto, becoming a regular at the neural computation workshops that Hinton organized in the late 2000s. Just as Google Brain was created, Andrew Ng brought him to the lab as one of its first hires. He was among the researchers who worked on the Cat Paper and the new Android speech service. Then, in the summer of 2013, Facebook came calling.

That year, Facebook hosted the Bay Area Vision Meeting, an annual gathering of computer vision researchers from across Silicon Valley. This mini-symposium was organized by Lubomir Bourdev, the Facebook engineer who urged Zuckerberg to buy DeepMind, and when a Facebook colleague suggested that Ranzato would be the perfect keynote speaker, Bourdev arranged to meet the young

Italian researcher for lunch at Google headquarters, about seven miles south of the Facebook campus along Highway 101. Initially, Ranzato assumed that Bourdev was angling for a job at Google, but as lunch progressed it became clear that not only did the Facebook engineer want Ranzato to address the Bay Area Vision Meeting, he also wanted him to join Facebook. Ranzato demurred. Though he wasn't completely happy at Google Brain—he spent more time on engineering work and less on the kind of creative research he preferred—Facebook didn't seem like an improvement. It didn't even have an AI lab. But over the next few weeks, with phone calls and emails, Bourdev kept asking.

At one point, Ranzato rang his old graduate school advisor, Yann LeCun, about the Facebook offer. LeCun did not approve. Back in 2002, he had been in a similar position. Google, then only four years old, had offered LeCun a job as its head of research, and he had turned the company down because he worried about its capacity for this kind of work. (At the time Google had only about six hundred employees.) "It was clear that Google was on a very good trajectory, but it wasn't at the scale where it could afford to do research," he says. Moreover, Google seemed to be focused more on short-term results than long-term planning. Many saw this as one of the company's great strengths. It was what allowed Google to deploy its deep learning speech engine on Android phones in only six months, racing past Microsoft and IBM and grabbing a market of considerable consequence. But such a focus on immediate results had concerned LeCun then, and it concerned him now that Facebook appeared to be operating in much the same way. "They don't do research," LeCun told Ranzato. "Make sure you can actually do research."

Still, Ranzato agreed to meet with Bourdev again, this time at Facebook headquarters, and near the end of their afternoon visit,

Bourdev said there was another person he'd like him to meet. They walked across campus and into another building, and as they approached a glass-walled conference room, there was Mark Zuckerberg. A few days later, Ranzato agreed to join the company. Promising to build a lab for long-term research, Zuckerberg gave Ranzato a desk beside his own. In the years that followed, this became a key part of the way Zuckerberg and Schroepfer pushed the company into new technological areas, from deep learning to virtual reality. Each new group sat beside the boss. In the beginning, this rubbed some at the company the wrong way. The rest of the Facebook brain trust felt that a long-term research lab planted next to Zuckerberg would clash with the company's move-fast-and-break-things culture and spread resentment among the rank and file. But at Facebook, Zuckerberg held sway. He was the founder and CEO, and unlike most chief executives, he controlled a majority of the voting shares on the board of directors.

A month later, Zuckerberg phoned Yann LeCun. He explained what the company was doing and asked for help. LeCun was flattered, particularly when Zuckerberg made a point of saying he'd read LeCun's research papers. But LeCun said he was happy as an academic in New York and that he couldn't do much more than provide some advice. "I can consult with you," he said. "But that's about it." He'd had similar conversations with Schrep in the past, and his stance had always been the same. Zuckerberg, though, kept pushing. Facebook was at another dead end. Schrep had approached several other leaders in the field, from Andrew Ng to Yoshua Bengio, and still the company didn't have anyone to run its lab—someone with the heft needed to attract the world's top researchers.

Then, in late November, Ranzato told Zuckerberg he was heading to NIPS. "What's NIPS?" Zuckerberg asked. Ranzato explained that hundreds of AI researchers were descending on a hotel and casino in Lake Tahoe, and Zuckerberg asked if he could tag along. Ranzato said this would be a little awkward, given that Zuckerberg was a pop-cultural icon, but he suggested they could avoid the distraction of Zuckerberg casually dropping into the conference unannounced if his boss arranged to give a speech in Lake Tahoe. So Zuckerberg arranged a speech with the conference organizers, then went a step further. Knowing that LeCun would be in Silicon Valley for a workshop the week before NIPS, he invited the NYU professor to dinner at his home in Palo Alto.

Zuckerberg lived in a white clapboard colonial-style home, tucked among the trees of the carefully manicured neighborhoods surrounding Stanford University. At dinner with LeCun, just between the two of them, Zuckerberg explained his grand vision for AI at Facebook. In the future, he told LeCun, interactions on the social network would be driven by technologies powerful enough to perform tasks on their own. In the short term, these technologies would identify faces in photos, recognize spoken commands, and translate between languages. In the longer term, "intelligent agents" or "bots" would patrol Facebook's digital world, take instructions, and carry them out as need be. Need an airline reservation? Ask a bot. Order flowers for your wife? A bot could do that, too. When LeCun asked if there were any areas of AI research that would not interest Facebook, Zuckerberg said: "Probably robotics." But everything else—everything in the digital realm—was in bounds.

The bigger issue was how Zuckerberg viewed the *philosophy* of corporate research. LeCun believed in "openness"—concepts, algorithms, and techniques openly shared with the wider community of researchers, not sequestered inside a single company or university. The idea was that this free exchange of information accelerated the progress of research as a whole. Everyone could build on the work of everyone else. Open research was the norm among academics in the field, but typically, the big Internet companies treated their most important technologies as trade secrets, jealously guarding the details from outsiders. Facebook, Zuckerberg explained, was the big exception. The company grew up in the age of open-source software—software code freely shared over the Internet—and it had extended the concept across the length and breadth of its technological empire. It even shared the designs of the custom-built hardware inside the massive computer data centers that served Facebook to the world. Zuckerberg believed that the value of Facebook lay in the people who used the social network, not in the software or hardware. Even with the raw materials, no one could re-create it, but if the company shared the raw materials, others could help improve them. LeCun and Zuckerberg were on common ground.

The next day, LeCun visited Facebook HQ, chatting with Zuckerberg, Schrep, and others in "The Aquarium," the glass-walled conference room where the Facebook boss held his meetings. At this point, Zuckerberg did not mince words. "We need you to build a Facebook AI lab," he said. LeCun said he had two conditions: "I am not moving from New York. And I am not leaving my job at NYU." Zuckerberg agreed to both—on the spot. Over the next few days, the company also hired Rob Fergus, another NYU professor, who had just won the next iteration of the ImageNet contest alongside a young graduate student named Matt Zeiler. Then Zuckerberg flew to NIPS. After unveiling his new lab at the private Facebook party the night before the conference began, he revealed the news to the rest of the world with his speech in the main hall.

WHEN Geoff Hinton sold his company to Google, he arranged to keep his professorship at the University of Toronto. He didn't want to abandon his students or leave what was now his home city. It was a unique arrangement. Previously, Google had always insisted that any academics it employed either take a leave of absence from their universities or quit entirely. But Hinton wouldn't accept this, even though the new arrangement was not exactly to his financial advantage. "I realized that the University of Toronto was paying me less than my pension would have been," he says. "So I was paying them to be allowed to teach." The largest expense for DNNresearch, Hinton's start-up, was the money he paid the lawyer who negotiated his agreement with Google—about $400,000. This contract set the template for LeCun and so many other academics who followed Hinton into industry. Much like Hinton, LeCun split his time between NYU and Facebook, though the ratio was very different. He spent one day a week at the university and four at the company.

Since most of the leading researchers at places like Google and Facebook came from academia—and so many remained academics, at least in part—Yann LeCun's vision of open research became the norm. "I don't know how to do research unless it's open, unless we are part of the research community," LeCun says. "Because if you do it in secret, you get bad-quality research. You can't attract the best. You're not going to have people who can push the state of the art." Even old hands, like Jeff Dean, who were brought up on a company culture of confidentiality, came to see the advantages of openness. Google began sharing its research as openly as Facebook or any other tech giant, publishing research papers describing its latest technologies and even open-sourcing much of its software.

This accelerated the development of these technologies, and it helped attract top researchers, which accelerated the process even more.

The loser in this brave new world was Microsoft. The company had seen the rise of deep learning firsthand when Hinton and his students joined forces with Li Deng on speech recognition, and the company had embraced the technology inside its speech labs in both the United States and China. In late 2012, after Google rolled its new speech engine onto Android phones, Rick Rashid, Microsoft's head of research, showed off the company's own speech work at an event in China, unveiling a prototype that could take in spoken words and translate them into another language. He liked to say that many of those watching in the audience wept when they saw and heard what the technology could do. Then, in the fall of 2013, longtime Microsoft vision researcher Larry Zitnick recruited Berkeley postdoc student Ross Girshick to build a new computer vision lab dedicated to deep learning. He'd been impressed by a talk Girshick had given in which he had described a system that pushed image recognition beyond what Hinton and his students demonstrated the previous December. Among those joining them was a young researcher named Meg Mitchell, who began applying similar techniques to language. Mitchell, a Southern Californian who had studied computational linguistics in Scotland, would later become a key figure in the deep learning movement after she told *Bloomberg News* that artificial intelligence suffered from a "sea of dudes" problem—that this new breed of technology would fall short of its promise because it was built almost entirely by men. It was an issue that would come to haunt the big Internet companies, including Microsoft. At the moment, these three researchers were working to build systems that could read photos and automatically generate captions. But although the lab tried to move with the cultural mood of the times—the team worked side by side at desks in a wide-open area of the office, a Silicon Valley–style arrangement that was still unusual inside Microsoft Research—progress was slow. Part of the problem was that they were training their neural networks on a few meager GPU machines tucked under their desks. The other part was that they were using the "wrong" software.

In the '90s, when the company dominated the worldwide software business, its primary source of strength had been the Windows operating system, a system that ran more than 90 percent of the world's home and business PCs and a majority of the servers that delivered online applications from inside the world's data centers. But by 2014, Microsoft's deep commitment to Windows was weighing the company down. The new wave of Internet businesses and computer scientists didn't use Windows. They used Linux, the open-source operating system that was free to use and modify. Linux provided a far cheaper and more flexible means of building the massively distributed systems that defined the Internet age, including deep learning. In fashioning these systems, the worldwide community of AI researchers freely traded all sorts of building blocks based on Linux, and these Microsoft researchers were stuck with Windows, spending much of their time just trying to find the next kludge that would allow these Linux tools to run on a Microsoft operating system.

So, when Facebook came calling, they left. Facebook offered the chance to build this new AI much quicker, to push it out into the marketplace much sooner, and, crucially, to plug into all the work already under way at Google and across so many other companies and academic labs. This was not an arms race like the one that Microsoft won in the '90s. It was an arms race in which companies gave away the arms—or at least many of them. Microsoft had seen what was happening, and then a competitor stole its edge.

Facebook hired both Girshick and Zitnick. Then Meg Mitchell went to Google.

The other challenge—and not just for Microsoft—was the huge expense of recruiting and keeping top researchers. Because the talent in this field was so rare—and its price had been set by Google's acquisitions of DNNresearch and DeepMind—the industry giants paid researchers millions or even tens of millions of dollars over a four- or five-year period, including salary, bonuses, and company stock. One year, according to DeepMind's annual financial accounts in Britain, its staff costs totaled $260 million for only seven hundred employees. That was $371,000 per employee. Even young PhDs, fresh out of grad school, could pull in half a million dollars a year, and the stars of the field could command more, in part because of their unique skills, but also because their names could attract others with the same skills. As Microsoft vice president Peter Lee told *Bloomberg Businessweek*, the cost of acquiring an AI researcher was akin to the cost of acquiring an NFL quarterback. Adding to this cutthroat atmosphere was the rise of another player. After Facebook unveiled its research lab and Google acquired DeepMind, it was announced that Andrew Ng would be running labs in both Silicon Valley and Beijing—for Baidu.

8

HYPE

"SUCCESS IS GUARANTEED."

In 2012, Alan Eustace was on a cross-country flight, reading one of those complimentary magazines you find in the back pocket of an airplane seat, when he came across a profile of the Austrian daredevil Felix Baumgartner. Baumgartner and his team were planning a one-man leap from the stratosphere, building a new kind of capsule that would carry the Austrian into the heavens as if he were an astronaut. But Eustace thought their approach was all wrong. They would be better off, he thought, if they treated Baumgartner not as an astronaut but as a scuba diver: he was sure that a scuba suit was ultimately a more nimble way of providing everything a human being would need to stay alive and well up in the thin air. Felix Baumgartner soon set a world skydiving record when he jumped from a capsule twenty-four miles above the Earth. But Eustace was already aiming to break the record. Over the next two years, he spent much of his spare time working with a private engineering firm to build a high-flying scuba suit and everything else he would need to top Baumgartner's mark. He planned on making the leap in the fall of 2014, from a spot miles above an abandoned runway in Roswell, New Mexico. But before he did, he took one last leap with Google.

After buying Krizhevsky, Sutskever, and Hinton for $44 million and DeepMind for $650 million, Eustace had nearly cornered the market in dyed-in-the-wool deep learning researchers. What the company still lacked was the hardware these researchers needed to accelerate their work in ways that matched both their talent and their ambition, as the trio from Toronto quickly discovered. Krizhevsky had won ImageNet with code written for GPU chips, but when they arrived in Mountain View, they found that Google's version, built by a researcher named Wojciech Zaremba, used standard chips, like everything else built for DistBelief, the company's bespoke hardware and software system for running neural networks. Hinton objected to the name of the project: it was called WojNet, after Zaremba, until Hinton started calling it AlexNet and the global community of AI researchers followed suit. Krizhevsky objected to Google's technology. The company had spent months building its system for running neural networks, but he had no interest in using it.

In his first days at the company, he went out and bought a GPU machine from a local electronics store, stuck it in the closet down the hall from his desk, plugged it into the network, and started training his neural networks on this lone piece of hardware. Other researchers shoved GPU machines under their desks. It wasn't all that different from the way Krizhevsky worked in his bedroom while still in Toronto, though Google was now paying for electricity. The rest of the company built and ran its software across the company's vast network of data centers, drawing on what may have been the world's largest private collection of computers, but Krizhevsky had to settle for something much smaller. The Googlers who controlled the company's data centers saw no reason to fill them with GPUs.

What these more conventionally minded Googlers didn't appreciate was that deep learning was the future—and that GPUs could accelerate this emerging technology at a rate ordinary computer chips could not. This often happened inside oversized tech companies, and small companies, too: Most people couldn't see beyond what they were already doing. The trick, Alan Eustace believed, was to surround yourself with people who could apply new kinds of expertise to problems that seemed unsolvable with the old techniques. "Most people look at particular problems from a particular point of view and a particular perspective and a particular history," he says. "They don't look at the intersections of expertise that will change the picture." It was the same philosophy he applied to his leap from the stratosphere. As he planned the jump, his wife didn't want him to take it. She insisted he shoot a video of himself explaining why he was taking the risk, so she could show it to their children if he didn't survive. He shot the video but told her the risk was minimal, almost nonexistent. He and his team had found a new way of making the leap, and though others might not understand it, he knew it would work. "People ask me a lot: 'Are you a daredevil?' But I am the opposite of a daredevil," he says. "I hire the best people I can find, and we all work together to basically eliminate every possible risk and test for every risk and try to get to the point where what is seemingly very dangerous is actually very safe."

Sitting in an office just down the hall from Krizhevsky, Jeff Dean knew Google's hardware needed to change. The company couldn't keep pushing the boundaries of deep learning unless it rebuilt DistBelief around GPUs. So in the spring of 2014, he met with Google's "head of artificial intelligence," John Giannandrea, known to everyone at the company as "J.G.," who oversaw both Google Brain and a sister team of AI specialists he helped build during these years. He was the person that researchers like Krizhevsky came to when they needed more GPUs under their desks or in the closet down the hall. He and Jeff Dean sat down to discuss how many graphics chips they should pack into a giant data center so that researchers like Krizhevsky wouldn't keep asking for more.

The first suggestion was twenty thousand. Then they decided this number was much too small. They should ask for forty thousand. But when they submitted their request to the Google bean counters, it was rejected out of hand. A network of forty thousand GPUs would cost the company about $130 million, and though Google regularly invested such enormous sums of money in its data center hardware, it had never invested in hardware like this. So Dean and Giannandrea took their request to Alan Eustace, who was about to make his leap from the stratosphere. Eustace understood. He took the request to Larry Page, and just before he broke Baumgartner's skydiving record in a scuba suit, he secured $130 million in graphic chips. Less than a month after the chips were installed, all forty thousand of them were running around the clock, training neural network after neural network after neural network.

BY then, Alex Krizhevsky was working for a completely different part of the company. That December, while visiting his parents back in Toronto during the holiday break, he'd received an email from a woman named Anelia Angelova, who wanted help with the Google self-driving car. She didn't actually work on the self-driving car. She worked with Krizhevsky at Google Brain. But she knew the lab's ongoing research in computer vision—an extension of Krizhevsky's efforts at the University of Toronto—would remake the way the company built its autonomous vehicles. The Google self-driving car project, known inside the company as Chauffeur, was nearly five years old. That meant Google had spent nearly five years building autonomous vehicles without help from deep learning.

At Carnegie Mellon in the late 1980s, Dean Pomerleau had designed a self-driving car with help from a neural network, but when Google went to work on autonomous vehicles nearly two decades later, the heart of the research community, including the many Carnegie Mellon researchers hired for the Google project, had long since discarded the idea. A neural network could help build a car that drove the empty streets on its own, but not much more. It was a curiosity, not a path to building vehicles that could navigate heavy traffic as human drivers do. Angelova, though, was not convinced. In an empty Google office building, after everyone else had gone home for the holidays, she started tinkering with deep learning as a way for cars to detect pedestrians as they crossed the street or strolled down the sidewalk. Because this was all very new to her, she reached out to the man she called "the master of deep networks." He agreed to help, and so, over the holiday break, she and Krizhevsky built a system that learned to identify pedestrians by analyzing thousands of street photos. When the company reopened after the New Year, they shared their new prototype with the leaders of the car project. It was so effective, they both were invited to work on Chauffeur, later rechristened Waymo after spinning off into its own company. Google Brain eventually gave Krizhevsky's desk to an intern because he almost never used it. He was always at Chauffeur.

The Chauffeur engineers called him "the AI whisperer," and his methods soon spread across the project. Deep learning became a way for the Google car to recognize *anything* on the road, from stop signs to street markings to other vehicles. Krizhevsky called it "low-hanging fruit." Over the next few years, he and his colleagues pushed the technology into other parts of the car's navigation system. Trained on the right data, deep learning could help plan a route forward or even predict future events. The team had spent the previous five years coding the car's behavior by hand. They could now build systems that learned behavior on their own. Rather than trying to define what a pedestrian looked like, one line of code at a time, they could train a system in a matter of days using thousands of streetside photos. In theory, if Google could gather enough data—images showing *every* scenario a car might encounter on the road—and then feed it into a giant neural network, this single system could do all the driving. That future was still many years away, at best. But in 2014, this was the direction Google turned.

The moment was part of a much larger shift inside the company. This single idea—a neural network—now transformed the way Google built technology across its growing empire, in both the physical world and the digital. With help from those forty thousand GPU chips and soon many more—a data center overhaul the company called Project Mack Truck—deep learning moved into everything from the Google Photos app, where it instantly found objects in a sea of images, to Gmail, where it helped predict the word you were about to type. It also greased the wheels inside AdWords, the online ad system that generated a vast majority of the company's $56 billion in annual revenue. By analyzing data showing which ads people had clicked on in the past, deep learning could help predict what they would click on in the future. More clicks meant more money. Google was spending hundreds of millions of dollars buying GPU chips—and millions more acquiring researchers—but it was already making those dollars back.

Soon, Amit Singhal, the head of Google Search, who had so vehemently resisted deep learning when approached by Andrew Ng and Sebastian Thrun in 2011, acknowledged that Internet technology was changing. He and his engineers had no choice but to relinquish their tight control over how their search engine was built. In 2015, they unveiled a system called RankBrain, which used neural networks to help choose search results. It helped drive about 15 percent of the company's search queries and was, on the whole, more accurate than veteran search engineers when trying to predict what people would click on. Months later, Singhal left the company after being accused of sexual harassment, and he was replaced as the head of Google Search by the head of artificial intelligence: John Giannandrea.

In London, Demis Hassabis soon revealed that DeepMind had built a system that reduced power consumption across Google's network of data centers, drawing on the same techniques the lab used to crack Breakout. This system decided when to turn on cooling fans inside individual computer servers and when to turn them off, when to open the data center windows for additional cooling and when to close them, when to use chillers and cooling towers and when the servers could get by without them. The Google data centers were so large and the DeepMind technology was so effective, Hassabis said, it was already saving the company hundreds of millions of dollars. In other words, it paid for the cost of acquiring DeepMind.

The power of Google's GPU cluster was that it allowed the company to experiment with myriad technologies on a massive scale. Building a neural network was a task of trial and error, and with tens of thousands of GPU chips at their disposal, researchers could explore more possibilities in less time. The same phenomenon quickly galvanized other companies. Spurred by the $130 million in graphics chips it sold to Google, Nvidia reorganized itself around the deep learning idea, and soon it was not merely selling chips for AI research, it was doing its own research, exploring the boundaries of image recognition and self-driving cars, hoping to expand the market even further. Led by Andrew Ng, Baidu explored everything from new ad systems to technology that could predict when the hard drives inside its data centers were about to fail. But the biggest change was the rise of the talking digital assistant. These services didn't just accept keywords typed into a Web browser and respond with a few Internet links, like a search engine. They could listen to your questions and commands and answer them audibly, as a person would. After Google remade speech recognition on Android phones, eclipsing what was possible with Apple Siri, the same technology spread across the industry. In 2014, Amazon unveiled Alexa, moving this technology from phones onto the living room coffee table, and the rest of the market quickly followed. Google's technology, now called Google Assistant, ran both on phones and on coffee-table devices. Baidu and Microsoft and even Facebook built their own assistants.

As all these products, services, and ideas proliferated and the marketing arms of these and so many other tech companies promoted them with the usual hyperbole, "artificial intelligence" became the buzz term of the decade, repeated ad infinitum across press releases, websites, blogs, and news stories. As always, it was a loaded term. For the general public, "artificial intelligence" revived the tropes of science fiction—conversational computers, sentient machines, anthropomorphic robots that could do anything a human could do and might end up destroying their creators. It didn't help that the press evoked films like *2001: A Space Odyssey* and *The Terminator* in the headlines, photos, and stories that sought to describe the new wave of technology. It was Frank Rosenblatt and the Perceptron all over again. As deep learning rose to the fore, so did the notion of a self-driving car. And at just about the same moment, a team of academics at the University of Oxford released a study predicting that automated technologies would soon cut a giant swath through the job market. Somehow, it was all mixed into one enormous, overflowing stew of very real technological advances, unfounded hype, wild predictions, and concern for the future. "Artificial intelligence" was the term that described it all.

The press needed heroes for its AI narrative. It chose Hinton, LeCun, Bengio, and sometimes Ng, thanks largely to the promotional efforts of Google and Facebook. The narrative did not extend to Jürgen Schmidhuber, the German researcher based on Lake Lugano who carried the torch for neural networks in Europe during the 1990s and 2000s. Some took issue with Schmidhuber's exclusion, including Schmidhuber. In 2005, he and Alex Graves, the researcher who later joined DeepMind, had published a paper describing a speech recognition system based on LSTMs—neural networks with short-term memory. "This is that crazy Schmidhuber stuff," Hinton had told himself, "but it actually works." Now the technology was powering speech services at companies like Google and Microsoft, and Schmidhuber wanted his due. After Hinton, LeCun, and Bengio published a paper on the rise of deep learning in *Nature*, he wrote a critique arguing that "the Canadians" weren't as influential as they seemed to be—that they'd built their work on the ideas of others working in Europe and Japan. And around the same time, when Ian Goodfellow presented his paper on GANs—a technology whose influence would soon reverberate across the industry—Schmidhuber stood up in the audience and chastised him for not citing similar work in Switzerland from the 1990s. He did this kind of thing so often, it became its own verb—as in: "You have been Schmidhubered." But he was hardly the only one working to claim credit for what was happening. After the community spent years ignoring their ideas, many deep learning researchers felt the urge to trumpet their own personal contributions to a very real technological revolution. "Everybody has a little Trump inside them," Hinton says. "You can see this in yourself, and it is good to be aware."

One exception was Alex Krizhevsky. As Hinton said: "He hasn't got enough Trump in him." Sitting at his desk inside Chauffeur, Krizhevsky was at the heart of this AI boom, but he didn't see his role as all that important, and he didn't see any of it as artificial intelligence. It was deep learning, and deep learning was just mathematics, pattern recognition, or, as he called it, "nonlinear regression." These techniques had been around for decades. It was merely that people like him had come along at the right time, when there was enough data and enough processing power to make it all work. The technologies he built were in no way intelligent. They worked only in very particular situations. "Deep learning should not be called AI," Krizhevsky says. "I went to grad school for curve fitting, not AI." What he did, first at Google Brain and then inside the self-driving car project, was apply the math to new situations. This was very different from any effort to re-create the brain—and far from vague fears that machines would someday spin outside our control. It was computer science. Others agreed, but this was not a view that made headlines. The louder voice belonged to his old Toronto labmate Ilya Sutskever.

IN 2011, while he was still at the University of Toronto, Sutskever flew to London for a job interview at DeepMind. He met with

Demis Hassabis and Shane Legg near Russell Square, and as the three men talked, Hassabis and Legg explained what they were trying to do. They were building AGI—artificial general intelligence—and they were beginning with systems that played games. As he listened, Sutskever thought they'd lost touch with reality. AGI was not something serious researchers talked about. So he turned down a job with the start-up and returned to the university, ultimately ending up at Google. But once inside the company, he realized that the very nature of AI research was changing. It was no longer about one or two people in an academic lab tinkering with a neural network. It was about big teams of people, all working toward big common goals, with massive amounts of computing power behind them. He was always one for big ideas, and as he moved into Google Brain, his ideas got bigger. After spending two months at the DeepMind offices as part of a transatlantic cooperation between the London lab and Google Brain, he came to believe that the only way to make real progress was to reach for what seemed like the unreachable. What he had in mind was different from the aims of Jeff Dean (who was more concerned with making an immediate impact on the market), and it was different from those of Yann LeCun (who was intent on looking into the future with his research, but never too far). It was closer to the views articulated by the founders of DeepMind. He talked as if the distant future was very close—machines that could outthink humans, computer data centers that could build other computer data centers. All that he and his colleagues needed was more data and more processing power. Then they could train a system to do *anything*—not just drive a car but read and talk and think. "He is somebody who is not afraid to believe," says Sergey Levine, a robotics researcher who worked alongside Sutskever at Google during these years. "There are many people who are not afraid, but he especially is not afraid."

By the time Sutskever joined Google, deep learning had remade both speech and image recognition. The next big step was *machine translation*—technology that could instantly translate any language into any other. This was a harder problem. It didn't involve identifying a single thing, like a dog in a photo. It involved taking a *sequence of things* (like the words that make up a sentence) and converting it into another sequence (the translation of that sentence). This would require a very different kind of neural network, but Sutskever believed a solution wasn't far away, and he wasn't alone. Two colleagues at Google Brain shared his vision. At places like Baidu and the University of Montreal, too, others were starting down the same path.

Google Brain had already explored a technology called "word embeddings." This involved using a neural network to build a mathematical map of the English language by analyzing a vast collection of text—news articles, Wikipedia articles, self-published books—to show the relationship between every word in the language and every other word. This wasn't a map you could ever hope to visualize. It didn't have two dimensions, like a road map, or three dimensions, like a video game. It had thousands of dimensions, like nothing you've ever seen, or could ever see. On this map, the word "Harvard" was close to the words "university," "Ivy," and "Boston," even though those words weren't linguistically related. The map gave each word a mathematical value that defined its relationship to the rest of the language. This was called a "vector." The vector for "Harvard" looked a lot like the vector for "Yale," but they weren't identical. "Yale" was close to "university" and "Ivy," but not "Boston."

Sutskever's translation system was an extension of this idea. Using the Long Short-Term Memory method developed by Jürgen Schmidhuber and Alex Graves in Switzerland, Sutskever fed reams of English text into a neural network *alongside* their French translations. By analyzing both the original text and the translation, this neural network learned to build a vector for an English sentence and then map it to a French sentence with a similar vector. Even if you knew no French, you could see the power of the math. The vector for "Mary admires John" was a lot like those for "Mary is in love with John" and "Mary respects John"—and very different from the one for "John admired Mary." The vector for "She gave me a card in the garden" matched those for "I was given a card by her in the garden" and "In the garden, she gave me a card." By the end of the year, the system built by Sutskever and his collaborators exceeded the performance of every other translation technology, at least with the small collection of English and French translations they were testing.

In December 2014, back at the NIPS conference, this time in Montreal, Sutskever presented a paper describing their work to a roomful of researchers from across the globe. The strength of the system, he told his audience, was its simplicity. "We use minimum innovation for maximum results," he said, as applause rippled across the crowd, catching even him by surprise. The power of a neural network, he explained, was that you could feed it data and it learned behavior on its own. Though training these mathematical systems was sometimes like black magic, this project was not. "It wanted to work," he said. It took in the data and trained for a while and gave results without all the usual trial and error. But Sutskever didn't see it merely as a breakthrough in translation. He saw it as a breakthrough for *any AI problem that involved a sequence*, from automatically generating captions for photos to instantly summarizing a news article with a sentence or two. Anything that a human could do in a fraction of a second, he said, a neural network could also do. It just needed the right data. "The real conclusion is that if you have a very large dataset and a very large neural network," he told his audience, "then success is guaranteed."

Geoff Hinton watched the speech from the back of the room. As Sutskever said "success is guaranteed," he thought: "Only Ilya could get away with that." Some researchers bristled at the audacity of the claim, but others were drawn to it. Somehow, Sutskever could say it without breeding too much resentment. It was who he was, and though it would be ridiculously boastful coming from someone else, it was somehow genuine when it came from him. He was also right—at least when it came to translation. Over the next eighteen months, Google Brain took this prototype and transformed it into a commercial system used by millions of people, an echo of what the lab had done with Navdeep Jaitly's speech prototype three years earlier. But here, the lab changed the equation in a way that would send another ripple across the field and, in the end, amplify the ambitions of Ilya Sutskever and many others.

"**WE** need another Google," Jeff Dean told Urs Hölzle, the Swiss-born computer scientist who oversaw the company's data centers. It was true. In the months after Google released its new speech recognition service on select Android phones, Dean realized there was a problem: If Google continued to expand the service and it eventually reached the more than 1 billion Android phones across the globe and those 1 billion phones used it just three minutes a day, the company would need twice as many data centers to handle all the extra traffic. That was a problem of epic proportions. Google already operated more than fifteen data centers—from California to Finland to Singapore—each built at a cost of hundreds of millions of dollars. During a standing meeting with Hölzle and a few other Googlers who specialized in data center infrastructure, however, Dean suggested an alternative: They could build a new computer chip just for delivering neural networks.

Google had a long history of building its own data center hardware. Its data centers were so large, sucking up so much electrical power, Hölzle and his team spent years designing computer servers, networking gear, and other equipment that could deliver Google services in cheaper and more efficient ways. This little-discussed enterprise undercut commercial hardware makers like HP, Dell, and Cisco and eventually took huge chunks out of their core business. As Google built its own hardware, it didn't need to buy on the open market, and as Facebook, Amazon, and others followed suit, these giants of the Internet created a shadow industry of computer hardware. But Google had never gone so far as to build its own computer chip, and neither had its rivals. That required an added level of expertise and investment that just didn't make economic sense. Companies like Intel and Nvidia produced chips on such an enormous scale, the cost was not something Google could match, and the chips did the job they needed to do. It was Nvidia's GPU chips that were powering the rise of deep learning, helping to train systems like the Android speech service. But now Dean was dealing with a new problem. After training the service, he needed a more efficient way of *running it*—serving it up across the Internet, delivering it to the world at large. Dean could deliver it with GPUs or standard processors, but neither was as efficient as he needed them to be. So he and his team built a new chip just for running neural networks. They gathered the funds from various other groups around Google, including the search group. By this time, everyone had seen what deep learning could do.

For years, Google designed data center hardware inside a semisecret lab in, of all places, Madison, Wisconsin. Hölzle—a former computer science professor with a diamond stud earring and an uplift of short salt-and-pepper hair—saw this work as the company's true competitive advantage, jealously guarding designs from the eyes of rivals like Facebook and Amazon. Madison was an out-of-the-way place that nevertheless attracted a steady stream of talent from the engineering school at the University of Wisconsin. Now Dean and Hölzle tapped into this talent for the new chip project while also hiring seasoned chip engineers from Silicon Valley companies like HP. The result was the tensor processing unit, or TPU. It was designed to process the tensors—mathematical objects—that underpinned a neural network. The trick was that its calculations were *less precise* than typical processors. The number of calculations made by a neural network was so vast, each calculation didn't have to be exact. It dealt in integers rather than floating point numbers. Rather than multiply 13.646 by 45.828, the TPU lopped off the decimal points and just multiplied 13 and 45. That meant it could perform trillions of extra calculations each second—exactly what Dean and his team needed, not only for the speech service but for translation.

What Sutskever had produced was research, not a product ready for mass consumption. His system worked well with common words but not with larger vocabularies, and it couldn't really compete with the existing translation service—built with good old-fashioned rules and statistics—that Google had offered over the Internet for more than a decade. But thanks to all the data he had gathered, the company had collected a vast array of translations that could help train a far larger neural network using the methods demonstrated by Sutskever and his fellow researchers. Their dataset was somewhere between a hundred and a thousand times larger than the one Sutskever had trained his system on. So, in 2015, Dean tapped three engineers to build a system that could learn from all that data.

Google's existing translation service broke sentences into pieces, converted them into slivers of another language, and then worked to connect these fragments into one coherent whole—hence late-night television host Jimmy Fallon's jokes about Google Translate's disjointed, slightly scrambled, less-than-coherent sentences. For English and French, its BLEU score, the standard way of measuring translation performance, was in the high twenties, which meant it didn't work very well. This represented an improvement of a little more than three points over four years. In just a few months of work, Dean's team built a neural network that topped the existing system by seven points. The method's great strength, as with all deep learning methods, was that this was a single learned task. There was no need to break the sentences into pieces. "Suddenly, things went from incomprehensible to comprehensible," says Macduff Hughes, who oversaw the team that built the old system. "It was like somebody turned the lights on."

But there was a problem. It took ten seconds to translate a ten-word sentence. That would never fly on the open Internet. People just wouldn't use it. Hughes thought the company would need three years to hone the system to the point where it could deliver translations without delay. Dean, however, thought otherwise. "We can do it by the end of the year, if we put our minds to it," he told Hughes during a company meeting at a hotel in San Francisco. Hughes was skeptical, but told his team to prepare for a new service by the end of the year. "I'm not going to be the one to say Jeff Dean can't deliver speed," he said.

They were racing against Baidu. The Chinese Internet giant had published a paper describing similar research several months earlier, and that summer, it published another, showing performance on par with the system built inside Google Brain. As Jeff Dean and his team built the new version of Google Translate, they decided the service would make its debut with English and Chinese. Because of the vast differences between the two languages, this was the pairing where deep learning provided the largest improvement. It was also the pairing where translation could, in the long run, provide the most benefit. These, after all, were the world's two largest economies. In the end, the Google engineers beat Dean's deadline by three months, and the difference was the TPU. A sentence that needed ten seconds to translate on ordinary hardware back in February could translate in milliseconds with help from the new Google chip. They released the first incarnation of the service just after Labor Day, well before Baidu. "I was amazed that it worked that well. I think everybody was," Hinton says. "Nobody expected it to work so well, so soon."

WHEN Geoff Hinton came to Google, he and Jeff Dean worked on a project they called "Distillation." This was a way of taking one of the giant neural networks they were training inside the company and then shrinking everything it had learned down to a size that allowed Google to actually use it in live online services, instantly delivering its skills to people around the globe. It was a marriage of Hinton's long career (neural networks) with Dean's (global computing). Then Hinton looked beyond neural networks, to a new and more complex effort to mimic the brain. It was an idea that first came to him in the late 1970s, and he called it a "capsule network." In the summer after Google acquired DeepMind, Hinton planned to spend three months at the London lab, and decided he would spend them working on this new old idea.

He bought two tickets on the *Queen Mary 2* from New York to Southampton, England—one for himself and one for his wife, Jackie Ford, the art historian he married in the late '90s, after his previous wife, Rosalind, died of ovarian cancer. They were scheduled to sail from New York on a Sunday. On the Thursday before they left Toronto, Jackie was diagnosed with advanced pancreatic cancer. The doctors gave her about a year to live, and advised her to start chemotherapy immediately. Knowing there was no chance of a cure, she decided to take the trip to Britain before starting her treatment when they returned to Toronto in the fall. Her family and many of her friends were still in England, and this would be her last chance to see them. So she and Hinton traveled to New York and left for Southampton on the Sunday. Hinton did spend the summer working on capsule networks, but he didn't make much progress.

9

ANTI-HYPE

"HE COULD PRODUCE SOMETHING EVIL BY ACCIDENT."

On November 14, 2014, Elon Musk posted a message on a website called Edge.org. Inside labs like DeepMind, he said, artificial intelligence was improving at an alarming rate:

> Unless you have direct exposure to groups like DeepMind, you have no idea how fast—it is growing at a pace close to exponential. The risk of something seriously dangerous happening is in the five-year time frame. Ten years at most. This is not a case of crying wolf about something I don't understand. I am not alone in thinking we should be worried. The leading AI companies have taken great steps to ensure safety. They recognise the danger, but believe that they can shape and control the digital superintelligences and prevent bad ones from escaping into the internet. That remains to be seen. . . .

Within the hour, the message vanished. But it wasn't all that different from what Musk had been saying for months, both in public and in private.

A year earlier, minutes after sitting down for dinner in Silicon Valley with *Bloomberg Businessweek* reporter Ashlee Vance, Musk said his big fear was that Larry Page was building an army of artificially intelligent robots that could end up destroying the human race. The trouble was not that Page was malicious. Page was a close friend. Musk often slept on his couch. The trouble was that Page operated under the assumption that anything Google built would do the world good. As Musk put it: "He could produce something evil by accident." The conversation stayed private for years, until Vance published his biography of Musk, but soon after their dinner, Musk was saying much the same thing across both national TV and social media. During an appearance on CNBC, he invoked *The Terminator.* "There have been movies about this," he said. On Twitter, he called artificial intelligence "potentially more dangerous than nukes."

The same tweet urged his followers to read *Superintelligence: Paths, Dangers, Strategies*, a recently published tome from an Oxford University philosopher named Nick Bostrom. Like Shane Legg, the founder of DeepMind, Bostrom believed that superintelligence could secure the future of humanity—or destroy it. "This is quite possibly the most important and most daunting challenge humanity has ever faced," he wrote. "And—whether we succeed or fail—it is probably the last challenge we will ever face." His concern was that scientists would design a system to perfect a particular part of our lives without realizing it would one day wreak havoc in ways no one had the power to stop. His oft-repeated metaphor was a system designed to create as many paperclips as possible. Such a system, he said, could transform "first all of earth and then increasing portions of space into paperclip manufacturing facilities."

That fall, Musk appeared onstage at a *Vanity Fair* conference in New York City, warning author Walter Isaacson about the dangers of artificial intelligence designed for "recursive self-improvement." If researchers designed a system to fight email spam, he explained, it could end up deciding that the best way of eliminating all the spam was just to remove all the people. When Isaacson asked if he would use his SpaceX rockets to escape these killer robots, Musk said escape might not be possible. "If there's some apocalypse scenario," he said, "it may follow people from Earth."

A few weeks later, Musk posted his message to Edge.org, a website overseen by an organization that explored new scientific ideas and hosted an annual gathering called the Billionaires' Dinner that included such luminaries as Musk, Larry Page, Sergey Brin, and Mark Zuckerberg, and was soon enveloped in controversy after the billionaire Jeffrey Epstein, one of its primary financial backers, was arrested for sex trafficking before killing himself in a jail cell. With his message on the Edge Foundation website, Musk was more explicit than he'd been in the past. He pointed to DeepMind as the evidence that the world was racing toward superintelligence. He said danger was five to ten years away at most. And as one of its investors, he had seen inside DeepMind before the London lab was suddenly acquired by Google. It was unclear what he had seen, if anything, that others had not.

Musk posted the message on a Friday. On the following Wednesday, he sat down for dinner with Mark Zuckerberg. It was the first time the two had met. Zuckerberg invited Musk to his white clapboard home under its leafy canopy in Palo Alto, hoping to convince the South African entrepreneur that all this talk about the dangers of superintelligence didn't make much sense. He had balked when the founders of DeepMind insisted they wouldn't sell their lab without a guarantee that an independent ethics board would oversee their AGI, and now, as Musk amplified this message across television and social media, he didn't want lawmakers and policy makers getting the impression that companies like Facebook would do the world harm with their sudden push into artificial intelligence. To help make his case, he also invited Yann LeCun, Mike Schroepfer, and Rob Fergus, the NYU professor who worked alongside LeCun at the new Facebook lab. The Facebookers spent the meal trying to explain that Musk's views on AI had been warped by a few misguided voices that were very much in the minority. The philosophical musings of Nick Bostrom, Zuckerberg and his fellow Facebookers said, were in no way related to what Musk had seen at DeepMind or inside any other AI lab. A neural network was still a long way from superintelligence. DeepMind built systems that optimized point totals inside games like Pong or Space Invaders, but they were useless elsewhere. You could shut the game down just as easily as you could a car.

But Musk was unmoved. The trouble, he said, was that AI was improving so quickly. The risk was that these technologies would cross the threshold from innocuous to dangerous before anyone realized what was happening. He laid down all the same arguments he made in his tweets and TV spots and public appearances, and as he talked, no one could quite tell if this was really what he believed or if he was just posturing, with an eye toward some other endgame. "I genuinely believe this is dangerous," he said.

DAYS after the dinner in Palo Alto, Elon Musk phoned Yann LeCun. He said he was building a self-driving car at Tesla, and he asked LeCun who he should hire to run the project. That week, he contacted several other Facebook researchers, asking each the same question—a gambit that eventually raised the ire of Mark Zuckerberg. LeCun told Musk he should contact Urs Muller, an old colleague from Bell Labs who'd built a start-up for exploring autonomous vehicles through deep learning. Before Musk could hire this Swiss researcher, however, someone else did. Days after LeCun got the call from Musk, he fielded an identical request from Jensen Huang, the founder and CEO of Nvidia, and he gave the same answer, which Nvidia acted on without delay. The company's ambition was to build a lab that would push the boundaries of self-driving and, in the process, help the company sell more GPU chips.

Even as Musk sounded the alarm that the race for artificial intelligence could destroy us all, he was joining it. For the moment, he was chasing the idea of a self-driving car, but he would soon chase the same grandiose idea as DeepMind, creating his own lab in pursuit of AGI. For Musk, it was all wrapped up in the same technological trend. First image recognition. Then translation. Then driverless cars. Then AGI.

He was part of a growing community of researchers, executives, and investors who warned against the dangers of superintelligence even as they were trying to build it. This included the founders and early backers of DeepMind as well as many of the thinkers drawn into their orbit. To outside experts, it seemed like nonsense. There was no evidence that superintelligence was anywhere close to reality. Current technologies were still struggling to reliably drive a car or carry on a conversation or just pass an eighth-grade science test. Even if AGI was close, the stance of people like Musk seemed like a contradiction. "If it was going to kill us all," the voices asked, "why build it?" But to those on the inside of this tiny community, it was only natural to consider the risks of what they saw as a uniquely important set of technologies. Someone was going to build superintelligence. It was best to build it while guarding against the unintended consequences.

Back in 2008, Shane Legg described this attitude in his thesis, arguing that although the risks were great, so were the potential rewards. "If there is ever to be something approaching absolute power, a super intelligent machine would come close. By definition, it would be capable of achieving a vast range of goals in a wide range of environments," he wrote. "If we carefully prepare for this possibility in advance, not only might we avert disaster, we might bring about an age of prosperity unlike anything seen before." He acknowledged that the attitude seemed extreme, but he also pointed to others who held similar beliefs. As they created DeepMind, he and Hassabis tapped into this community. They came to Peter Thiel through the Singularity Summit. They secured another investment from Jaan Tallinn, one of the founders of the Skype Internet phone-calling service, who would soon join a group of academics in creating what they called the Future of Life Institute, an organization dedicated to exploring the existential risks of artificial intelligence and other technologies. Then Hassabis and Legg carried these ideas to new places. They introduced them to Musk, and they took them to Facebook and Google as the two tech giants scrambled to acquire their young start-up. As they wooed investors and buyers, Legg wasn't shy about his view of the future. Superintelligence would arrive in the next decade, he would say, and so would the risks. Mark Zuckerberg balked at these ideas—he merely wanted DeepMind's talent—but Larry Page and Google embraced them. Once they were inside Google, Suleyman and Legg built a DeepMind team dedicated to what they called "AI safety," an effort to ensure that the lab's technologies did no harm. "If technologies are going to be successfully used in the future, the moral responsibilities are going to have to be baked into their design by default," Suleyman says. "One has to be thinking about the ethical considerations the moment you start building the system." As Elon Musk invested in DeepMind and began to express many of the same concerns and launched his own efforts in the field, he was joining a movement. Then he took it to extremes.

The Future of Life Institute was less than a year old in the fall of 2014 when it invited this growing community to a private summit in Puerto Rico. Led by an MIT cosmologist and physicist named Max Tegmark, it aimed to create a meeting of the minds along the lines of the Asilomar conference, a seminal 1975 gathering where the world's leading geneticists discussed whether their work—gene editing—would end up destroying humanity. The invitation the institute sent included two photos: one showing the beach in San Juan, the other showing some poor soul shoveling through a snowdrift that had buried a Volkswagen Beetle somewhere in colder climes (meaning: "At the beginning of January, you will be much happier in Puerto Rico"). It also promised the event would not include the press (meaning: "You can freely discuss your concerns about the future of artificial intelligence without waking up to headlines about *The Terminator*"). It called this closed-door gathering "The Future of AI: Opportunities and Challenges." Demis Hassabis and Shane Legg attended. So did Elon Musk. On the first Sunday of 2015, six weeks after his dinner with Mark Zuckerberg, Musk took the stage to discuss the threat of an intelligence explosion, the moment when artificial intelligence suddenly reaches a level that even the experts haven't anticipated. That, he said, was the big risk: The technology could suddenly cross over into the danger area without anyone quite realizing it. This was an echo of Bostrom, who was also onstage in Puerto Rico, but Musk had a way of amplifying the message.

Jaan Tallinn had seeded the Future of Life Institute with a pledge of $100,000 a year. In Puerto Rico, Musk pledged $10 million, earmarked for projects that explored AI safety. But as he prepared to announce this new gift, he had second thoughts, fearing the news would detract from an upcoming launch of a SpaceX rocket and its landing on a drone ship in the Pacific Ocean. Someone reminded him there were no reporters at the conference and that the attendees were under Chatham House Rules, meaning they'd agreed not to reveal what was said in Puerto Rico, but he was still wary. So he made the announcement without mentioning the dollar figure. A few days later, after his rocket crashed during its attempted landing, he revealed the $10 million grant in a tweet. For Musk, the threat of superintelligence was only one thing among many. His main concern, it seemed, was maximum attention. "He is a super-busy man, and he doesn't have time to dig into the nuances of the issues, but he understands the basic outlines of the problem," Tallinn says. "He also genuinely enjoys the press attention, which translates to his very slogan-y tweets, et cetera. There is a symbiosis between Elon and the press that annoys many AI researchers, and that is the price the community has to pay."

At the conference, Tegmark distributed an open letter that sought to codify the common beliefs of those who gathered in Puerto Rico. "We believe that research on how to make AI systems robust and beneficial is both important and timely," the letter read, before recommending everything from labor market forecasts to the development of tools that could ensure AI technology was safe and reliable. Tegmark sent a copy to each attendee, giving all the opportunity to sign. The tone of the letter was measured and the contents straightforward, sticking mostly to matters of common sense, but it served as a marker for those who were committed to the idea of AI safety—and were at least willing to listen to the deep concerns of people like Legg, Tallinn, and Musk. One person who attended the conference but did not sign was Kent Walker, Google's chief legal officer, who was more of an observer than a participant in Puerto Rico as his company sought to expand its AI efforts both in California with Google Brain and in London with DeepMind. But most of the other attendees signed, including one of the top researchers inside Google Brain: Ilya Sutskever.

Max Tegmark later wrote a book about the potential impact of superintelligence on the human race and the universe as a whole. In the opening pages, he described a meeting between Elon Musk and Larry Page at a dinner party held in the wake of the Puerto Rico conference. After food and cocktails somewhere in California's Napa Valley, Page mounted a defense of what Tegmark described as "digital utopianism"—"that digital life is the natural and desirable next step in the cosmic evolution and that if we let digital minds be free rather than try to stop or enslave them, the outcome is almost certain to be good." Page worried that paranoia over the rise of AI would delay this digital utopia, even though it had the power to bring life to worlds well beyond the Earth. Musk pushed back, asking how Page could be sure this superintelligence wouldn't end up destroying humanity. Page accused him of being "specieist" because he favored carbon-based life-forms over the needs of new species built with silicon. At least for Tegmark, this debate over late-night cocktails showed the attitudes that were lining up against each other in the heart of the tech industry.

ABOUT six months after the conference in Puerto Rico, Greg Brockman walked down Sand Hill Road, the short stretch of asphalt that winds past more than fifty of the biggest venture capital firms in Silicon Valley. He was headed for the Rosewood—the upscale, urban, California ranch–style hotel where entrepreneurs plotted their pitches to the big venture capitalists—and he couldn't stop worrying about the time. After stepping down as chief technology officer at Stripe, a high-profile online payments start-up, the twenty-six-year-old MIT dropout was on his way to dinner with Elon Musk, and he was late. But when Brockman walked into the private dining room at the Rosewood, Musk had yet to arrive, and, in typical fashion, the founder and CEO of Tesla and SpaceX didn't arrive for more than an hour. But another conspicuous Silicon Valley investor was already there: Sam Altman, president of the start-up accelerator Y Combinator. He greeted Brockman and introduced him to the small group of AI researchers who'd gathered on the patio that looked out onto the hills west of Palo Alto. One of them was Ilya Sutskever.

After they moved inside and sat down for dinner, Musk arrived, filling the room with his unusually broad shoulders and equally expansive personality. But he, like the rest of them, wasn't quite sure what they were all doing there. Altman had brought them together in the hopes of building a new AI lab that could serve as a counterweight to the labs that were rapidly expanding inside the big Internet companies, but no one knew if that was possible. Brockman certainly wanted to build one after leaving Stripe, one of Y Combinator's most successful companies. He had never actually worked in AI and had only recently bought his first GPU machine and trained his first neural network, but as he told Altman a few weeks earlier, he was intent on joining the new movement. So, too, was Musk, after watching the rise of deep learning inside Google and DeepMind. But no one was sure how they could enter a field that was already dominated by the richest companies in Silicon Valley. So much of the talent had begun making enormous sums of money inside Google and Facebook, not to mention Baidu, newly invigorated after poaching Andrew Ng as chief scientist, and

Twitter, which had just acquired two notable deep learning startups. Altman had invited Sutskever and a few other like-minded researchers to the Rosewood to help explore the possibilities, but they spent the evening asking questions rather than giving answers. "There was this big question: Is it too late to start a lab with a bunch of the best people? Is that a thing that can happen? No one could quite say it was impossible," Brockman remembers. "There were people who said: 'It's really hard. You need to get this critical mass. You need to work with the best people. How are you going to do that? There's chickens and eggs here.' The thing that I heard was that it's not impossible."

When he drove home with Altman that night, Brockman vowed to build the new lab they all seemed to want. He began with calls to several leaders in the field, including Yoshua Bengio, the University of Montreal professor who helped foster the deep learning movement alongside Geoff Hinton and Yann LeCun, but he made it clear that he was still committed to academia. Bengio drew up a list of promising young researchers from across the community, and as Brockman contacted these and many others, he grabbed the interest of several researchers who shared at least some of Musk's concern over the dangers of artificial intelligence. Five of them, including Ilya Sutskever, had recently spent time inside DeepMind. They were drawn to the idea of a lab outside the control of the big Internet companies and completely detached from the profit motive that drove them. That, they believed, was the best way of ensuring that artificial intelligence progressed in a safe way. "Very few scientists think about the long-term consequences of their work," says Wojciech Zaremba, who was among the researchers approached by Brockman. "I wanted the lab to take seriously the possibility that AI might have sweeping negative implications to the world despite being an incredibly fulfilling intellectual puzzle."

But none of these researchers would commit to a new lab unless the others did. To break the deadlock, Brockman invited his top ten choices to spend an autumn afternoon at a winery in Napa Valley, north of San Francisco. The group included Sutskever and Zaremba, who had moved to Facebook after his stint at Google. Brockman hired a bus to take them from his San Francisco apartment into wine country, and even that, he felt, helped solidify their big idea. "An underrated way to bring people together are these times where there is no way to speed up getting to where you're going," he says. "You have to get there, and you have to talk."

In Napa, they discussed a new kind of virtual world, a digital playground where AI software agents could learn to do anything on a personal computer that a human could do. This would drive the development of DeepMind-style reinforcement learning not just inside games like Breakout but inside *any software application*, from Web browsers to Microsoft Word. This, they believed, was a path to truly intelligent machines. After all, a Web browser extended to the entire Internet. It was a gateway to any machine and any person. To navigate a web browser, you needed not just motor skills but language skills. This was a task that would stretch the resources of even the largest tech companies, but they resolved to tackle it without a company behind them. They envisioned a lab that was entirely free of corporate pressures, a not-for-profit that would give away all its research, so that anyone could compete with the Googles and the Facebooks. By the end of the weekend, Brockman invited all ten researchers to join this lab, and he gave them three weeks to think it over. Three weeks later, nine out of ten were on board. They called their new lab OpenAI. "It felt like the right level of drastic-ness," Sutskever says. "I enjoy doing the most drastic thing possible. This felt like the most drastic thing possible."

But before they could unveil the lab to the rest of the world, researchers like Zaremba and Sutskever would have to tell Facebook and Google. In addition to his work at Google Brain and the Facebook AI lab, Zaremba had spent time at DeepMind. After he agreed to join OpenAI, the giants of the Internet offered what he called "borderline-crazy" amounts of money to change his mind—two or three times his market value. And these offers were tiny compared to the figures Google threw at Sutskever—many millions of dollars a year. Both declined the offers, but then bigger offers arrived, even as they flew to Montreal for NIPS, where they were set to unveil the new OpenAI lab. A conference that once attracted a few hundred researchers now spanned nearly four thousand, with bodies filling the lecture halls where the top thinkers presented the top papers and countless companies scrambling to arrange meetings in the side rooms as they fought for the most prized tech talent on the planet. It was like a western mining town during the gold rush.

Once he arrived in Montreal, Sutskever met with Jeff Dean, who made another offer in an effort to keep him at Google. He couldn't help but consider it. Google offered two or three times what OpenAI was going to pay him, which was nearly $2 million for the first year. Musk and Altman and Brockman had no choice but to delay their announcement while they waited for Sutskever to decide. He phoned his parents back in Toronto, and as he continued to weigh the pros and cons, Brockman sent text after text urging him to choose OpenAI. This went on for days. Finally, on Friday, the last day of the conference, Brockman and the others decided they needed to announce the lab with or without him. The announcement was set for three p.m., and that time came and went, without an announcement and without a decision from Sutskever. Then he texted Brockman to say he was in.

Musk and Altman painted OpenAI as a counterweight to the

dangers presented by the big Internet companies. While Google, Facebook, and Microsoft still kept some technology under wraps, OpenAI—a not-for-profit backed by over a billion dollars in funding promises from Musk, Peter Thiel, and others—would give away the technology of the future without *any* restraint. AI would be available to everyone, not just the richest companies on Earth. Yes, Musk and Altman admitted, if they open-sourced all their research, the bad actors could use it as well as the good. If they built artificial intelligence that could be used as a weapon, anyone could use it as a weapon. But they argued that the threat of malicious AI would be mitigated precisely because their technology would be available to anyone. "We think it's far more likely that many, many AIs will work to stop the occasional bad actors," Altman said. It was an idealistic vision that would eventually prove to be completely impractical, but it was what they believed. And their researchers believed it, too. Whether or not their grand vision was viable, Musk and Altman were at least moving closer to the center of what seemed to be the world's most promising technological movement. Many of the top researchers were now working for them. Zaremba, who studied with Yann LeCun at NYU, said those "borderline-crazy" offers didn't tempt him. They turned him off and pushed him closer to OpenAI. He felt those big dollars were an effort not just to keep his services but to prevent the creation of the new lab. Sutskever felt much the same.

Not everyone bought the idealism preached by Musk, Altman, and the rest. At DeepMind, Hassabis and Legg were incensed, feeling betrayed not only by Musk, who had invested in their company, but by many of the researchers hired at OpenAI. Five of them had spent time at DeepMind, and for Hassabis and Legg, the new lab would create an unhealthy competition on the path to intelligent machines that could have dangerous consequences. If labs were racing each other to new technologies, they were less likely to realize where they might go wrong. In the coming months, Hassabis and Legg said as much to both Sutskever and Brockman. In the hours after OpenAI was unveiled, Sutskever heard even harsher words after he walked into a party at the conference hotel. The party was thrown by Facebook, and before the evening was out, he was approached by Yann LeCun.

Standing near the elevator in the corner of a wide-open space in the hotel lobby, LeCun told Sutskever he was making a mistake and gave more than ten reasons why. OpenAI's researchers were too young. The lab didn't have the experience of someone like himself. It didn't have the money of a company like Google or Facebook, and its not-for-profit structure wouldn't bring that money in. It had attracted a few good researchers, but it couldn't compete for talent in the long term. The idea that the lab would openly share all its research was not the draw it seemed to be. Facebook was already sharing most of its work with the larger community—and Google was beginning to do much the same. "You," LeCun told Sutskever, "are going to fail."

10

EXPLOSION

"HE RAN ALPHAGO LIKE OPPENHEIMER RAN THE MANHATTAN PROJECT."

On October 31, 2015, Facebook chief technology officer Mike Schroepfer stood at the end of a table inside the company's Disneyland of a corporate headquarters, addressing a roomful of reporters. Pointing at a slideshow splashed across a flat-panel display on the wall, he described the company's latest array of research projects—experiments with drones, satellites, virtual reality, AI. As was often the case with these carefully scripted events, most of it was old news. Then he mentioned that several Facebook researchers, spanning offices in New York and California, were teaching neural networks to play Go. Over the decades, machines had beaten the world's best players at checkers, chess, backgammon, Othello, even *Jeopardy!* But Go was the contest of human intellect no machine could crack. Just recently, *Wired* magazine had published a feature story about a French computer scientist who'd spent a decade trying to build AI that could challenge the world's best Go players. Like most everyone across the international community of AI researchers, he believed it would be another decade before he—or anyone else—reached those heights. But as Schrep told that roomful of reporters, Facebook researchers were confident they could crack the game much sooner using deep learning, and if they did, it would mark a major leap forward in artificial intelligence.

Go matches two players against each other on a nineteen-by-nineteen grid. They take turns placing stones at the intersections, trying to enclose portions of the board and, in the process, capture each other's pieces. Chess mimics a ground battle. Go plays out like the Cold War. A move in one corner of the board ripples everywhere else, changing the landscape of the game in subtle and often surprising ways. In chess, at each turn, there are about thirty-five possible moves to choose from. In Go, there are two hundred. It is exponentially more complex than chess, and in the mid-2010s that meant no machine, no matter how powerful, could calculate the outcome of every available move in any reasonable amount of time. But as Schroepfer explained, deep learning promised to change the equation. After analyzing millions of faces in millions of photos, a neural network could learn to distinguish you from your brother, your college roommate from everyone else. In much the same way, he said, Facebook researchers could build a machine that mimicked the skills of a professional Go player. By feeding millions of Go moves into a neural network, they could teach it to recognize what a good move looks like. "The best players end up looking at visual patterns, looking at the visuals of the board to help them understand what are good and bad configurations in an intuitive way," he explained. "So, we're using the patterns on the board—a visual rec system—to tune the possible moves the system can make."

On one level, Facebook was simply teaching machines to play a game, he said. On another level, in doing so, it was pushing toward

AI that could remake Facebook. Deep learning was refining the way ads were targeted on the company's social network. It was analyzing photos and generating captions for the visually impaired. It was driving Facebook M, the smartphone digital assistant under development inside the company. Using the same techniques that underpinned their Go experiments, Facebook researchers were building systems that aimed not just to recognize spoken words but to truly understand natural language. One team had recently built a system that could read passages from *The Lord of the Rings* and then answer questions about the Tolkien trilogy—complex questions, Schrep explained, that involved spatial relationships among people, places, and things. He also said that years would pass before the company's tech could crack Go, and certainly before it could truly understand natural language—but the path to these two futures was in place. It was a path that computer scientists had worked to build for decades, with much bluster and only small amounts of practical technology. Now, he said, the AI movement was finally catching up with its big ideas.

What he didn't tell those reporters was that others were moving down the same path. Days after news stories appeared describing Facebook's efforts to crack Go, one of those companies responded. Demis Hassabis appeared in an online video, looking straight into the camera, dominating the frame. It was a rare appearance from the founder of DeepMind. The London lab did most of its talking through research papers published in high-profile academic journals like *Science* and *Nature*, and typically spoke to the outside world only after a major breakthrough. In the video, Hassabis hinted at work still gestating inside the lab, work involving the game of Go. "I can't talk about it yet," he said, "but in a few months, I think there will be quite a big surprise." Facebook's bid for press attention had stirred its biggest rival. A few weeks after Hassabis appeared in that online video, a reporter asked Yann LeCun if DeepMind had possibly built a system that could beat a top Go player. "No," he said. And he said it more than once, partly because he thought the task was too difficult, but also because he hadn't heard anything. The community was that small. "If DeepMind had beaten a top Go player," LeCun said, "someone would have told me." He was wrong.

Days later, *Nature* carried a cover story in which Hassabis and DeepMind revealed that their AI system, AlphaGo, had beaten the three-time European Go champion. It had happened in a closed-door match in October. LeCun and Facebook caught wind of this the day before the news was announced. That afternoon, in a bizarre and hapless bid for preemptive PR personally driven by Zuckerberg, the company alerted the press to online posts from both Zuckerberg and LeCun boasting of Facebook's own Go research and the path this was building to other forms of AI inside the company. But the fact remained that Google and DeepMind were ahead. In that closed-door match, AlphaGo had won all five games against the European champion, a Chinese-Frenchman named Fan Hui. Several weeks later, in Seoul, it would challenge Lee Sedol, the world's best player of the last decade.

WEEKS after Google acquired DeepMind, Demis Hassabis and several other DeepMind researchers flew to Northern California for a powwow with the leaders of their new parent company and a demo of the lab's deep learning success with Breakout. Once the meeting was out of the way, they broke up into informal groups and Hassabis found himself chatting with Sergey Brin. As they talked, they realized they had a common interest: Go. Brin said that when he and Page were building Google at Stanford, he played so much Go that Page worried their company would never happen.

Hassabis said that if he and his team wanted to, they could build a system capable of beating the world champion. "I thought that was impossible," Brin said. In that moment, Hassabis resolved to do it.

Geoff Hinton compared Demis Hassabis to Robert Oppenheimer, the man whose stewardship of the Manhattan Project during the Second World War led to the first atomic bomb. Oppenheimer was a world-class physicist: He understood the science of the massive task at hand. But he also had the skills needed to motivate the sprawling team of scientists that worked under him, to combine their disparate strengths to feed the larger project, and to somehow accommodate their foibles as well. He knew how to move men (and women, including Geoff Hinton's cousin, Joan Hinton). Hinton saw the same combination of skills in Hassabis. "He ran AlphaGo like Oppenheimer ran the Manhattan Project. If anybody else had run it," Hinton says, "they would not have gotten it working so fast, so well."

David Silver, the researcher who had known Hassabis since Cambridge, and a second DeepMind researcher, Aja Huang, were already working on a Go project, and they soon joined forces with Ilya Sutskever and a Google intern named Chris Maddison, who had started their own project in Northern California. The four researchers published a paper on their early work around the middle of 2014, before the project expanded into a much larger effort, culminating in the victory over Fan Hui, the European Go champion, the next year. The result shocked both the worldwide Go community and the global community of AI researchers, but AlphaGo versus Lee Sedol promised to be something far bigger. When IBM's Deep Blue supercomputer topped world-class champion Garry Kasparov inside a high-rise on the West Side of Manhattan in 1997, it was a milestone for computer science, and it was widely and enthusiastically covered by the global press. But it was a tiny event compared to the match in Seoul. In Korea—not to mention Japan and China—Go is a national pastime. Over 200 million people would watch AlphaGo versus Lee Sedol, double the audience for the Super Bowl.

At a press conference the day before the five-game match, Lee boasted that he would win with ease: four games to one or even five games to none. Most Go players agreed. Though AlphaGo had beaten Fan Hui in a way that left no doubt the machine was the better player, there was a chasm between Fan Hui and Lee Sedol. ELO ratings, a relative measure of their talents, put Lee in an entirely different echelon of the game. But Hassabis believed the outcome would be very different. When he sat down for lunch with several reporters the following afternoon, two hours before the first game, he carried a copy of the *Korea Herald*, the country's peach-colored English-language daily. Both he and Lee Sedol were pictured on the front page, above the fold. He hadn't expected quite so much attention. "I expected it to be big," said the thirty-nine-year-old Englishman, boyishly small and balding on top. "But not that big." Still, over this lunch of dumplings and kimchi and grilled meats—which he didn't eat—Hassabis said he was "cautiously confident." What the pundits didn't grasp, he explained, was that AlphaGo had continued to hone its skills since the match in October. He and his team originally taught the machine to play Go by feeding 30 million moves into a deep neural network. From there, AlphaGo played game after game against itself, all the while carefully tracking which moves proved successful and which didn't—much like the systems the lab had built to play old Atari games. In the months since beating Fan Hui, the machine had played itself several million more times. AlphaGo was continuing to teach itself the game, learning at a faster rate than any human ever could.

Google chairman Eric Schmidt sat across from Hassabis during this pregame meal on the top floor of the Four Seasons, expounding on the merits of deep learning in his haughty way. At one point, someone called him an engineer. He corrected them. "I am not an engineer," he said. "I am a computer scientist." He recalled that when he had trained as a computer scientist in the 1970s, AI had seemed to carry so much promise, but that as the '80s and then the '90s arrived, it had never quite delivered. Now the promise was being realized. "This technology," he said, "is tantalizingly powerful." AI wasn't just a way of juggling photos, he said. It represented the future of Google's $75 billion Internet business, as well as countless other industries, including healthcare. Later, as they gathered several floors below to watch the game, Hassabis and Schmidt were joined by Jeff Dean. The mere presence of Schmidt and Dean showed how important the match was for Google. Three days later, as the match reached its climax, Sergey Brin flew into Seoul.

Hassabis spent the first game moving between a private viewing room and the AlphaGo control room down the hall. This room was filled with PCs and laptops and flat-panel displays, all tapping into a service running across several hundred computers inside Google data centers on the other side of the Pacific. The week before, a team of Google engineers had run their own ultra-high-speed fiber-optic cable into this control room to ensure a reliable connection to the Internet. As it turned out, the control room didn't need to provide much control: After several months of training, AlphaGo was able to play entirely on its own, with no human help. Not that Hassabis and his team could have helped even if they wanted to. None of them played Go at the level of grandmaster. All they could do was watch. "I can't tell you how tense it is," Silver said. "It's hard to know what to believe. You're listening to the commentators on the one hand. And you're looking at AlphaGo's evaluation on the other hand. And all the commentators are disagreeing."

On that first day of the match, alongside Schmidt and Dean and other Google VIPs, they watched the machine win. At the postgame press conference, sitting in front of hundreds of reporters and photographers from both East and West, Lee Sedol told the world he was in shock. "I didn't think that AlphaGo would play the game in such a perfect manner," the thirty-three-year-old said, through an interpreter. Over more than four hours of play, the machine proved it could match the talents of the best player in the world. Lee said that AlphaGo's talents had caught him off guard. He would change his approach in game two.

About an hour into game two, Lee Sedol stood up, left the game room, and walked out on a private patio for a smoke. While he was gone, Aja Huang, the Taiwan-born DeepMind researcher who sat across from Lee Sedol in the match room and physically made each move on behalf of AlphaGo, placed a black stone in a largely empty space on the right-hand side of the board, beside and a little below a single white stone. It was the thirty-seventh move of the game. In the commentary room around the corner, Michael Redmond, the only Western Go player to reach nine dan, the game's highest rank, did a double take. "I don't really know if it's a good move or a bad move," Redmond told the more than 2 million English speakers following the match online. His co-commentator, Chris Garlock, longtime editor of an online magazine devoted to Go and the vice president of the American Go Association, said: "I thought it was a mistake." Lee returned after several minutes and spent several more just staring at the board. In all, he took nearly fifteen minutes to respond, a huge chunk of the two hours he was allotted for play in the first phase of the game—and he never quite regained his footing. More than four hours later, he resigned. He was down two games to nil.

Move 37 had also surprised Fan Hui, the man who had been so comprehensively beaten by the machine a few months earlier and who had since joined the DeepMind team to serve as AlphaGo's sparring partner in the run-up to the match with Lee Sedol. He had never managed to beat DeepMind's AI, but his encounters with AlphaGo had opened his eyes to new ways of playing. In fact, in the weeks after his defeat he had gone on a six-game winning streak against top (human) competition, his world ranking climbing to new heights in the process. Now, standing outside the commentary room on the seventh floor of the Four Seasons, in the few minutes following Move 37, he came to see the effect of this strange move. "It's not a human move. I've never seen a human play this move," he said. "So beautiful." He kept repeating the word. Beautiful. Beautiful. Beautiful.

The next morning, David Silver slipped into the control room, just so he could revisit the decisions AlphaGo made in choosing Move 37. In the midst of each game, drawing on its training with tens of millions of human moves, AlphaGo calculated the probability that a human would make a particular play. With Move 37, the probability was one in ten thousand. AlphaGo knew this wasn't a move a professional Go player would ever make. Yet it made the move anyway, drawing on the millions of games it had played with itself—games in which no human was involved. It had come to realize that although no human would make the move, it was still the right one. "It discovered this for itself," Silver said, "through its own process of introspection."

It was a bittersweet moment. Even as Fan Hui hailed the beauty of the move, a sadness fell over the Four Seasons and, indeed, all of Korea. On his way to the postgame press conference, a Chinese journalist named Fred Zhou ran into a *Wired* magazine reporter who had flown to Korea from the United States. Zhou said how happy he was to speak with another writer who cared about technology, complaining that other journalists were treating the event like sport. What they should cover, he said, was the AI. But then his tone changed. Though he was elated when AlphaGo won the first game, Zhou said, he now felt a deep despair. He pounded on his chest to show what he meant. The next day, Oh-hyoung Kwon, a Korean who runs a start-up incubator on the other side of Seoul, said that he, too, felt a sadness. This wasn't because Lee Sedol was Korean but because he was human. "There was an inflection point for all human beings," Kwon said, as several of his colleagues nodded in agreement. "It made us realize that AI is really near us—and realize the dangers of it, too." The somber mood only deepened over the weekend. Lee Sedol lost the third game and, thus, the match. Sitting onstage at the postgame press conference, the Korean was repentant. "I don't know what to say today, but I think I will have to express my apologies first," he said. "I should have shown a better result, a better outcome, a better contest." Minutes later, after apparently realizing he should be gracious in technical defeat, Mark Zuckerberg posted a message to Facebook congratulating Demis Hassabis and DeepMind. Yann LeCun did the same. But sitting next to Lee Sedol, Hassabis found himself hoping the Korean would win at least one of the two remaining games.

Seventy-seven moves into the fourth game, Lee froze again. It was a repeat of game two, only this time he took even longer to find his next move. The center of the board was packed with stones, both black and white, and for nearly twenty minutes, he stared at these stones, gripping the back of his neck and rocking back and forth. Finally, he placed his white stone between two black stones in the middle of the grid, effectively cutting the board in two. AlphaGo went into a tailspin. As each game progressed, AlphaGo would perpetually recalculate its chances of winning, posting a percentage on a flat-panel display in the control room. After Lee's move—Move 78—the machine responded with a move so poor, its chances of winning immediately plummeted. "All the thinking that AlphaGo had done up to that point was sort of rendered useless," Hassabis said. "It had to restart." At that moment, Lee looked up from the board and stared at Huang, as if he had gotten the better of the man, not the machine. From there, the machine's odds continued to drop, and after nearly five hours of play, it resigned.

Two days later, as he walked through the lobby of the Four Seasons, Hassabis explained the machine's collapse. AlphaGo had assumed that no human would ever make Move 78. It calculated the odds at one in ten thousand—a very familiar number. Like AlphaGo before him, Lee Sedol had reached a new level, and he said as much during a private meeting with Hassabis on the last day of the match. The Korean said that playing the machine had not only rekindled his passion for Go but opened his mind, giving him new ideas. "I have improved already," he told Hassabis, an echo of what Fan Hui had said several days earlier. Lee Sedol would go on to win his next nine matches against top human players.

The match between AlphaGo and Lee Sedol was the moment when the new movement in artificial intelligence exploded into the public consciousness. It was a milestone moment not just for AI researchers and tech companies but for people on the street. This was true in the United States, and even more so in Korea and China, just because, in these countries, Go was viewed as the pinnacle of intellectual achievement. The match revealed both the power of the technology and the fears that it would one day eclipse humanity, before reaching a moment of optimism, underscoring the often surprising ways the technology could push humanity to new heights. Even as Elon Musk warned of the dangers, it was a period of extreme promise for AI. After reading about the match, Jordi Ensign, a forty-five-year-old computer programmer from Florida, went out and got two tattoos. AlphaGo's Move 37 was tattooed on the inside of her right arm—and Lee Sedol's Move 78 was on the left.

11

EXPANSION

"GEORGE WIPED OUT THE WHOLE FIELD WITHOUT EVEN KNOWING ITS NAME."

The Aravind Eye Hospital sits at the southern tip of India, in the middle of a sprawling, crowded, ancient city called Madurai. Each day, more than two thousand people stream into this timeworn building, traveling from across India and sometimes other parts of the world. The hospital offers eyecare to anyone who walks through the front door, with or without an appointment, whether they can pay for care or not. On any given morning, dozens crowd into the waiting rooms on the fourth floor, as dozens more line up in the hallways, all waiting to walk into a tiny office where lab-coated technicians capture images of the backs of their eyes. This is a way of identifying signs of diabetic blindness. In India, nearly 70 million people are diabetic, and all are at risk of blindness. The condition is called diabetic retinopathy, and if detected early enough, it can be treated and stopped. Each year, hospitals like the Aravind scan millions of eyes, and then doctors examine each scan, looking for the tiny lesions, hemorrhages, and subtle discolorations that anticipate blindness.

The trouble is that India doesn't produce enough doctors. For every 1 million people, there are only eleven ophthalmologists, and in rural areas, the ratio is even smaller. Most people never receive the screening they need. But in 2015, a Google engineer named Varun Gulshan hoped to change that. Born in India and educated at Oxford before joining a Silicon Valley start-up that was acquired by Google, he officially worked on a virtual-reality gadget called Google Cardboard. But in his "20 percent time," he started exploring diabetic retinopathy. His idea was to build a deep learning system that could automatically screen people for the condition, without help from a doctor, and so identify far more people that needed care than doctors ever could on their own. He soon contacted the Aravind Eye Hospital, and it agreed to share the thousands of digital eye scans that he would need to train his system.

Gulshan didn't understand how to read these scans himself. He was a computer scientist, not a doctor. So he and his boss roped in a trained physician and biomedical engineer named Lily Peng, who happened to be working on the Google search engine. Others had tried to build systems for automatically reading eye scans in the past, but these efforts had never matched the skills of a trained physician. The difference this time was that Gulshan and Peng were using deep learning. Feeding thousands of retinal scans from the Aravind Eye Hospital into a neural network, they taught it to recognize signs of diabetic blindness. Such was their success, Jeff Dean pulled them into the Google Brain lab, around the same time that DeepMind was tackling Go. The joke among Peng and the rest of her medically minded team was that they were a cancer that metastasized into the Brain. It wasn't a very good joke. But it was not a bad analogy.

THREE years earlier, in the summer of 2012, Merck & Co., one of the world's largest drug companies, launched a contest on a website called Kaggle. Kaggle was a place where any company could set up a contest for computer scientists, offering prize money to anyone who could solve a problem it needed solved. Offering a $40,000 prize, Merck provided a sprawling collection of data describing the behavior of a particular set of molecules and asked contestants to predict how they would interact with other molecules in the human body. The aim was to find ways of accelerating the development of new medicines. Two hundred thirty-six teams entered the contest, which was scheduled to run for two months. When George Dahl, Geoff Hinton's student, discovered the contest while riding on a train from Seattle to Portland, he decided to enter. He had no experience with drug discovery, just as he had no experience with speech recognition before building a system that shifted the future of the entire field. He also suspected that Hinton wouldn't approve of him entering the contest. But then Hinton liked to say he wanted his students working on something he wouldn't approve of. "It is sort of like the Gödel completeness result. What if he approves of you doing things he doesn't approve of? Is that really disapproval?" Dahl says. "Geoff understands the limits of his own abilities. He has some intellectual humility. He is open to surprises, to possibilities."

When Dahl returned to Toronto, he met with Hinton, and when Hinton asked, "What are you working on?," Dahl told him about Merck.

"I was on the train going to Portland and I just trained a really dumb neural net on this Merck data and I hardly did anything at all and it's already in seventh place," Dahl said.

"How much longer is the contest?" Hinton asked.

"Two weeks," Dahl said.

"Well," Hinton replied, "you have to win it."

Dahl wasn't really sure he could win it. He hadn't given the project all that much thought. But Hinton was insistent. This was the heady time between the success of deep learning with speech technology and the triumph with Go, and Hinton was keen to show just how adaptable neural networks could be. He now called them dreadnets (a play on the early-twentieth-century battleships called dreadnoughts), convinced they would sweep everything before them. Dahl was reminded of an old Russian joke Ilya Sutskever liked to tell in which the Soviet army runs out of shells while firing on its capitalist enemy. "What do you mean you're out of shells?" the Soviet general says to the sergeant who alerts him to the problem. "You're a communist!" So the army keeps on firing. Hinton said they had to win the contest, so Dahl enlisted the help of Navdeep Jaitly and a few other deep learning researchers from their lab in Toronto—and they won.

The contest explored a drug discovery technique called the quantitative structure-activity relationship, or QSAR, which Dahl had never heard of when he went to work on the Merck data. As Hinton put it: "George wiped out the whole field without even knowing its name." Soon, Merck added this method to the long and winding processes necessary to discover new medicines. "You can think of AI as a large math problem where it sees patterns that humans can't see," says Eric Schmidt, the former Google chief executive and chairman. "With a lot of science and biology, there are patterns that exist that humans can't see, and when pointed out, they will allow us to develop better drugs, better solutions."

In the wake of Dahl's success, countless companies took aim at the larger field of drug discovery. Many were start-ups, including a San Francisco company founded by one of George Dahl's labmates at the University of Toronto. Others were pharmaceutical giants like Merck that at least made loud noises about how this work would fundamentally change their business. All were years away from overhauling the field completely, if only because the task of drug discovery is prodigiously difficult and time-consuming. Dahl's discovery amounted to a tweak rather than a transforming breakthrough. But the potential of neural networks quickly galvanized research across the medical field.

When Ilya Sutskever published the paper that remade machine translation—known as the Sequence to Sequence paper—he said it was not really about translation. When Jeff Dean and Greg Corrado read it, they agreed. They decided it was an ideal way of analyzing healthcare records. If researchers fed years of old medical records into the same kind of neural network, they decided, it could learn to recognize signs that illness was on the way. "If you line up the medical records data, it looks like a sequence you're trying to predict," Dean says. "Given a patient in this particular stage, how likely are they to develop diabetes in the next 12 months? If I discharge them from the hospital, will they come back in a week?" He and Corrado soon built a team inside Google Brain to explore the idea.

It was in this environment that Lily Peng's blindness project took off—to the point where a dedicated healthcare unit was established inside the lab. Peng and her team acquired about one hundred and thirty thousand digital eye scans from the Aravind Eye Hospital and various other sources, and they asked about fifty-five American ophthalmologists to label them—to identify which included those tiny lesions and hemorrhages that indicated diabetic blindness was on the way. After that, they fed these images into a neural network. And then it learned to recognize telltale signs on its own. In the fall of 2016, with a paper in the *Journal of the American Medical Association*, the team unveiled a system that could identify signs of diabetic blindness as accurately as trained doctors, correctly spotting the condition more than 90 percent of the time, which exceeded the National Institutes of Health's recommended standard of at least 80 percent. Peng and her team acknowledged that the technology would have to clear many regulatory and logistical hurdles in the years to come, but it was ready for clinical trials.

They ran one trial at the Aravind Eye Hospital. In the short term, Google's system could help the hospital deal with the constant stream of patients moving through its doors. But the hope was that Aravind would also deploy the technology across the network of the more than forty "vision centers" it operated in rural areas around the country where few if any eye doctors were available. Aravind was founded in the late 1970s by a man named Govindappa Venkataswamy, an iconic figure known across India as "Dr. V." He envisioned a nationwide network of hospitals and vision centers that operated like McDonald's franchises, systematically reproducing inexpensive forms of eyecare for people across the country. Google's technology could play right into this idea—if they could actually get it in place. Deploying this technology was not like deploying a website or a smartphone app. The task was largely a matter of persuasion, not only in India but in the U.S. and the UK, where many others were exploring similar technology. The widespread concern among healthcare specialists and regulators was that a neural network was a "black box." Unlike with past technologies, hospitals wouldn't have the means to explain why a diagnosis was made. Some researchers argued that new technologies could be built to solve this issue. But it was a far from trivial problem. "Don't believe anyone who says that it is," Geoff Hinton told the *New Yorker* in a sweeping feature story on the rise of deep learning in healthcare.

Still, Hinton believed that as Google continued its work with diabetic retinopathy and others explored systems for reading X-rays, MRIs, and other medical scans, deep learning would fundamentally change the industry. "I think that if you work as a radiologist you are like Wile E. Coyote in the cartoon," he said during a lecture at a Toronto hospital. "You're already over the edge of the cliff, but you haven't yet looked down. There's no ground underneath." He argued that neural networks would eclipse the skills of trained doctors because they would continue to improve as researchers fed them more data, and that the black-box problem was something people would learn to live with. The trick was convincing the world it was not a problem, and this would come through testing—proof that even if you could not see inside of them, they did what they were supposed to do.

Hinton believed that machines, working alongside doctors, would eventually provide a hitherto impossible level of healthcare. In the near term, he argued, these algorithms would read X-rays, CAT scans, and MRIs. As time went on, they would also make pathological diagnoses, reading Pap smears, identifying heart murmurs, and predicting relapses in psychiatric conditions. "There's much more to learn here," Hinton told a reporter as he let out a small sigh. "Early and accurate diagnosis is not a trivial problem. We can do better. Why not let machines help us?" This was particularly important to him, he said, because his wife had been diagnosed with pancreatic cancer after it advanced beyond the stage where she could be cured.

IN the wake of AlphaGo's win in Korea, many inside Google Brain grew to resent DeepMind, and a fundamental divide developed between the two labs. Led by Jeff Dean, Google Brain was intent on building technologies with practical and immediate impact: speech recognition, image recognition, translation, healthcare.

DeepMind's stated mission was artificial general intelligence, and it was chasing this North Star by teaching systems to play games. Google Brain was an integral part of Google that delivered revenue. DeepMind was an independent lab with its own set of rules. At Google's new offices near St. Pancras station in London, DeepMind was cordoned off in its own section of the building. Its employees could get into the Google section with their company badges, but Google employees could not get onto the DeepMind floors. The separation grew sharper after Larry Page and Sergey Brin spun off several Google projects into their own businesses and moved them all under a new umbrella company called Alphabet. DeepMind was among those that became its own entity. The tension between Google Brain and DeepMind was so great, the two labs held a kind of summit behind closed doors in Northern California in an effort to ease the situation.

Mustafa Suleyman was one of the founders of DeepMind, but he seemed like a better fit for Google Brain. The man everyone called "Moose" wanted to build technology for today, not for the distant future. He wasn't a gamer or a neuroscientist or even an AI researcher. The son of a Syria-born London cabdriver, he was an Oxford dropout who created a helpline for Muslim youths and worked for the mayor of London on human rights. Nor was he the decidedly nerdish, often introverted type who typically worked in AI. More imposing than Hassabis and Legg, with a head of dark curls, a close-cropped beard, and a stud in his left ear, he was a hipster who prided himself on knowing all the best bars and restaurants, whether in London or New York, and he delivered his opinions loudly, without apology. When Elon Musk celebrated his fortieth birthday aboard the Orient Express, Suleyman was the DeepMind founder that joined this rolling Bacchanalia. He liked to say that when he and Hassabis were growing up together in

North London, he was not the nerdy one. They had not been close friends. Suleyman later remembered that in their younger years, when he and Hassabis discussed how they would change the world, they found little common ground. Hassabis would propose complex simulations of the global financial system that could solve the world's biggest social problems sometime in the distant future, and Suleyman would stop at the present. "We have to engage with the real world today," he would say. Some DeepMind employees felt that Suleyman was deeply jealous and resentful of Hassabis and Legg because they were scientists and he was not, that he was intent on proving he was just as important to DeepMind as they were. One colleague couldn't quite believe they had all founded the same company.

Like many inside Google Brain, Suleyman would come to resent AlphaGo. But in the beginning, the warm glow from DeepMind's Go-playing machine gave an added shine to his own pet project. Three weeks after DeepMind revealed that AlphaGo had beaten Fan Hui, the European champion, Suleyman unveiled what he called DeepMind Health. While he was growing up in London near King's Cross, his mother had been a nurse with the National Health Service, the seventy-year-old government institution that provided free healthcare to all British residents. Now his aim was to build artificial intelligence that could remake the world's healthcare providers, beginning with the NHS. Every news story covering the project pointed to AlphaGo as evidence that DeepMind knew what it was doing.

His first big project was a system for predicting acute kidney injury. Each year, one out of every five people admitted to a hospital develop the condition, when their kidneys suddenly stop working as they should, becoming unable to properly remove toxins from the bloodstream. Sometimes this permanently damages the kidneys.

Other times it leads to death. But if the condition is identified soon enough, it can be treated and stopped and reversed. With DeepMind Health, Suleyman aimed to build a system that could anticipate acute kidney disease by analyzing a patient's health records, including blood tests, vital signs, and past medical history. To do that, he needed data.

Before unveiling the new project, DeepMind signed an agreement with the Royal Free London NHS Foundation Trust, a government trust that operated several British hospitals. This provided patient data that DeepMind researchers could feed into a neural network so it could identify patterns that anticipated acute kidney injury. After the project was unveiled, AlphaGo went to Korea and defeated Lee Sedol, and the warm glow of the Go-playing machine grew brighter. Then, just a few weeks later, *New Scientist* magazine revealed the agreement between DeepMind and the Royal Free NHS Trust, showing just how much data was being shared with the lab. The deal gave DeepMind access to healthcare records for 1.6 million patients as they moved through three London hospitals as well as records from the previous five years, including information describing drug overdoses, abortions, HIV tests, pathology tests, radiology scans, and information about their particular hospital visits. DeepMind was required to delete it after the deal ended, but in Britain, a country that places a particularly high value on digital privacy, the story raised a specter that would follow DeepMind Health and Mustafa Suleyman for years. The following July, a British regulator ruled that the Royal Free NHS Trust had illegally shared its data with DeepMind.

12

DREAMLAND

"IT IS NOT THAT PEOPLE AT GOOGLE DRINK DIFFERENT WATERS."

In the spring of 2016, Qi Lu sat on a bicycle, rolling through the park in downtown Bellevue. The city's glass towers loomed overhead as he teetered down the promenade, struggling to keep his bicycle upright. It was no ordinary bicycle. When he turned the handlebars left, it moved right, and when he turned right, it moved left. He called it a "backwards brain bike," because the only way to ride was to think backwards. Conventional wisdom said: "You never forget how to ride a bicycle." But that is exactly what he hoped to do. Decades after he first learned to ride as a child growing up in Shanghai, he aimed to erase everything he'd learned and burn entirely new behavior into his brain. This, he believed, would show his company the way forward.

Lu worked for Microsoft. After joining the company in 2009, he oversaw the creation of Bing, its multibillion-dollar answer to the Google search engine monopoly. Seven years later, when he and his backwards bicycle wobbled through the park in downtown Bellevue, ten miles east of Seattle, just down the road from Microsoft headquarters, he was one of the company's most powerful executives, leading its latest push into artificial intelligence. But Microsoft was playing catch-up. The trouble, he knew only too well, was that the company had spent years struggling to make headway with new technology in new markets. For nearly ten years, it had battled for a place in the smartphone market, redesigning its Windows operating system to compete with the Apple iPhone and a world of Google Android phones, building a talking digital assistant that could challenge the speech technologies emerging from Google Brain, and spending no less than $7.6 billion to acquire Nokia, a company with decades of experience designing and selling mobile phones. But none of it had worked. The company's phones still felt like old-fashioned PCs, and in the end they captured almost none of the market. Microsoft's problem, Lu thought, was that it handled new tasks in old ways. It designed, deployed, and promoted technology for a market that no longer existed. After reading a series of essays from a Harvard Business School professor that deconstructed the foibles of aging corporations, he came to see Microsoft as a company still driven by the procedural memory burned into the brains of its engineers, executives, and middle managers when they first learned the computer business in the '80s and '90s, before the rise of the Internet, smartphones, open-source software, and artificial intelligence. The company needed to change its way of thinking, and with his backwards bicycle, Lu hoped to show that it could.

The bike was built by a Microsoft colleague named Bill Buxton and his friend Jane Courage. When Lu took this counterintuitive contraption on its first test run, they came with him. As Lu made his way across the park in downtown Bellevue—a tiny man with short black hair and wire-rimmed glasses, rolling past the shade trees and the reflecting pond and the waterfall—Buxton and Courage held up their iPhones, capturing the ride on video, one from the front, the other from the back. The idea was to share the experience with the rest of his Microsoft executive team, prove it could be done, and eventually get them on the bike, too—all thirty-five of them—so they could feel what it was like to change their thinking in this fundamental way. Lu knew it would take weeks to learn this new bike—and he knew that once he did, he would no longer have the memory needed to ride an ordinary bicycle. But his example, he hoped, would push Microsoft into the future.

After struggling to keep the bike upright for nearly twenty minutes, he set off down the promenade one last time. Then, as he turned the handlebars on his backwards bicycle, he fell and broke his hip.

FOUR years earlier, in the fall of 2012, Li Deng sat at his desk inside Building 99, the heart of the Microsoft Research lab, reading an unpublished paper describing the sweeping hardware and software system the new Google Brain lab used to train its neural networks. This was the system Google called DistBelief, and as part of a small committee reviewing papers for inclusion in the upcoming NIPS conference, Deng was laying eyes on the blueprints weeks before the rest of the world. After bringing Geoff Hinton and his students to the Microsoft Research lab, where they built a neural network that could recognize spoken words with unprecedented accuracy, Deng had watched from afar as Google beat Microsoft to market with the same technology. Now he realized this technology would spread well beyond spoken words. "When I read the paper," Deng remembers, "I realized what Google was doing."

Microsoft had spent more than twenty years investing in artificial intelligence, paying big money for many of the world's top researchers—and this put the company at a disadvantage as deep learning rose to the fore. Over the decades, the worldwide community of researchers had divided itself into distinct philosophical factions. In his history of artificial intelligence, *The Master Algorithm*, University of Washington professor Pedro Domingos called them "tribes." Each tribe nurtured its own philosophy—and often looked down on the philosophies of others. The connectionists, who believed in deep learning, were one tribe. The symbolists, who believed in the symbolic methods championed by the likes of Marvin Minsky, were another. Other tribes believed in ideas ranging from statistical analysis to "evolutionary algorithms" that mimicked natural selection. Microsoft had invested in AI at a time when the connectionists were not the top researchers. It hired from other tribes, and this meant that even as deep learning began to succeed in ways other technologies did not, many of the company's leading researchers harbored a deep prejudice against the idea of a neural network. "To be honest, the entire upper chain of Microsoft Research did not believe in it," Qi Lu says. "That was the environment."

Qi Lu wasn't the only one to worry about the entrenched culture at Microsoft. Hinton, too, had major reservations. He questioned the way Microsoft researchers, unlike, say, those at Google, worked on their own, insulated from any pressure to commercialize. "When I was an academic, I thought that was great, because you don't have to dirty your hands with development," Hinton says. "But in terms of actually getting the technology to a billion people, Google is far more efficient." He was also concerned by an article in *Vanity Fair* titled "Microsoft's Lost Decade," which explored the ten-year tenure of chief executive Steve Ballmer through the eyes of current and former Microsoft executives. One of the story's big revelations was the way Ballmer's Microsoft used a technique called "stack ranking" to review the performance of its employees and weed out a certain proportion of them regardless of their actual performance and promise. After Microsoft pulled out of the bid for Hinton's start-up, he told Deng he could never have joined such a company. "It wasn't the money. It was the review system," he said. "It may be good for salespeople. But it's not for researchers."

In any case, many at Microsoft were skeptical of deep learning. The company's vice president of research, Peter Lee, had seen deep learning remake speech recognition inside his own lab after Li Deng brought Geoff Hinton to Redmond, and he still didn't believe. That breakthrough seemed like a one-off. He had no reason to think the same technology would succeed with other areas of research. Then he flew to Snowbird, Utah, for a meeting of the U.S. computer science department chairs. Though he had stepped down as the chair of the computer science department at Carnegie Mellon, Lee still attended this annual confab as a way of keeping up with the latest academic trends, and that year, in Utah, he saw Jeff Dean give a speech on deep learning. When he returned, he arranged a meeting with Deng in a small conference room in Building 99 and asked him to explain what Dean was so excited about. Deng started to describe the DistBelief paper and what it said about Google's larger ambitions, explaining that Microsoft's chief rival was building infrastructure for a new future. "They are spending big money," he said. But Lee cut him off, knowing that under the rules of the NIPS conference, Deng wasn't allowed to discuss the paper until it was published. "That's an academic paper," he told Deng. "You can't show that to me." Deng didn't mention the paper again, but he kept talking about Google and Microsoft and where the technology was moving. At the end of it all, Lee still thought Google's ambition was misplaced. Speech recognition was one thing, image recognition was another, and both were only a small part of what any machine needed to do. "I just needed to know what is going on," he said. But he soon asked Deng to join a meeting of the lab's leading thinkers.

They gathered at another building across campus, in a much larger room. Deng stood at the lectern, in front of two dozen researchers and executives, his laptop plugged into a flat screen mounted on the wall behind him, ready to punctuate each big thought with a graph or a diagram or a photo. But as he started to explain the rise of deep learning, from the speech work at Microsoft to its spread across the industry, he was interrupted by a voice from across the room. It was Paul Viola, one of the company's leading experts in computer vision. "Neural networks have never worked," he said. Deng acknowledged the complaint and returned to his presentation. But Viola interrupted again, rising from his seat, walking to the front of the room, unplugging Deng's laptop from the flat panel on the wall, and attaching his own. Up on the screen, a book cover appeared, mostly orange, with swirls of purple and a one-word title printed in small white letters. It was Marvin Minsky's *Perceptrons.* Decades ago, Viola said, Minsky and Papert had proven that neural networks were fundamentally flawed and would never reach the lofty peaks so many had promised. Eventually, Deng continued his presentation—but Viola continued to interrupt. He interrupted so many times, a voice from across the room soon told him to keep quiet. "Is this Li's presentation or yours?" the voice said. It was Qi Lu.

If Qi Lu was a prime example of the cosmopolitan nature of the

AI community, his background made him one of the community's most unlikely participants. Brought up by his grandfather in a poverty-stricken countryside at the height of Mao Zedong's Cultural Revolution, he ate meat just once a year, when his family celebrated the spring festival, and attended a school where a single teacher taught four hundred students. Yet he overcame all the natural disadvantages he faced to take a degree in computing science at Shanghai's Fudan University and to attract the attention, in the late '80s, of the American computer scientist Edmund Clarke, who happened to be in China looking for talent he could bring back to his own university, Carnegie Mellon. Clarke gave a speech at Fudan on a Sunday, when Lu typically rode his bike across the city to visit his parents, but when a heavy rain came down, he stayed home. That afternoon, someone knocked on his door, urging him to fill a seat at Clarke's lecture. Too many seats were empty because of the rain. So Lu saw the lecture, and after impressing Clarke with his questions from across the auditorium, he was invited to apply for a spot at Carnegie Mellon. "It was luck," he remembers. "If it hadn't rained, I would have gone to see my parents."

When Lu joined the Carnegie Mellon PhD program, he knew very little English. One of his professors was Peter Lee, his future Microsoft colleague. During Lu's first year, Lee gave his class an exam that asked for code that could find the shortest path to the restroom from any spot in the Carnegie Mellon computer science building when "nature calls." In the middle of the exam, Qi Lu walked up to Lee and asked: "What is a nature call?" he said. "That's not a procedure I've ever heard of." Despite the language gap, it was obvious to Lee that Lu was a computer scientist of extreme and unusual talent. After Carnegie Mellon, Lu rose through the ranks at Yahoo! and then at Microsoft. By the time Li Deng gave his presentation in Building 99, Lu oversaw both the Bing search engine and various other parts of the company, working closely with Microsoft Research.

He saw himself as the rare technology executive who understood the technology, a strategist as well as a systems architect, a visionary who read the research papers emerging from the world's leading labs. He had a way of delivering his ideas in sharp, self-contained, slightly strange technological axioms:

Computing is the intentional manipulation of information toward a purpose.

Data is becoming the primary means of production.

Deep learning is computation on a new substrate.

Even before the meeting in Building 99, he knew where the industry was headed. Like Peter Lee, he had recently attended a private gathering of computer scientists where one of the founders of Google Brain trumpeted the rise of deep learning. At Foo Camp, an annual Silicon Valley gathering billed as an "unconference" where the attendees set the agenda as it goes along, he was part of a group that gathered around Andrew Ng as he explained the ideas behind the Cat Paper. At Microsoft, Lu was aware of the new speech technology that emerged after Hinton and his students visited the company, but it wasn't until he met Ng that he fully realized what was happening. His Bing engineers painstakingly built each piece of the Microsoft search engine by hand. But as Ng explained it, they could now build systems that learned these pieces on their own. In the weeks that followed, in typical fashion, he began reading the research literature emerging from places like New York University and the University of Toronto. The day Deng gave his presentation on the rise of deep learning, Lu listened, and he asked the right questions. So Deng knew what to do when Geoff Hinton emailed him a few weeks later, revealing the $12 million offer from Baidu. He forwarded the note to Qi Lu, and it was Lu who urged the leaders of Microsoft Research to join the auction for Hinton and his students. The leaders of Microsoft Research, however, were still skeptical.

WHEN Qi Lu returned to work several months after breaking his hip in the Bellevue park, he was still walking with a cane. Meanwhile, AlphaGo had beaten Lee Sedol, and the tech industry was stricken with a kind of AI fever. Even smaller Silicon Valley companies—Nvidia, Twitter, Uber—were jockeying for position in a race toward a single idea. After Twitter acquired Madbits, the company founded by Clément Farabet, the NYU researcher who turned down Facebook, Uber bought a start-up called Geometric Intelligence, a collection of academics pulled together by an NYU psychologist named Gary Marcus. Deep learning and deep learning researchers were the coin of the day. But Microsoft was handicapped. It wasn't an Internet company or a smartphone company or a self-driving car company. It didn't actually build the stuff that needed the Next Big Thing in artificial intelligence.

As he recovered from the first surgery on his hip, Lu urged the Microsoft brain trust to embrace the idea of a driverless car. Myriad tech companies and carmakers had a long head start with their autonomous vehicles, and Lu wasn't exactly sure how Microsoft would enter this increasingly crowded market. But that wasn't the issue. His argument wasn't that Microsoft should sell a driverless car. It was that Microsoft should *build* one. This would give the company the skills and the technologies and the insight it needed to succeed in so many other areas. Google had come to dominate so many markets, Lu believed, because it built a search engine in the era when the Internet was expanding as no one had ever seen. Engineers like Jeff Dean were forced to build technologies no one had ever built, and in the years that followed, these technologies drove everything from Gmail to YouTube to Android. "It is not that people at Google drink different waters," he said. "The search engine required them to solve a set of technological challenges." Building a self-driving car, Lu believed, would enrich Microsoft's future in the same way. "We must put ourselves in a position to see the future of computing."

The idea was ridiculous, but no more ridiculous than the ideas that drove Microsoft's biggest rivals. Google paying $44 million for Hinton and his students was "ridiculous." Just months later, as the rest of the market was throwing far higher sums at others in the field, it seemed like good business. In Korea, AlphaGo seemed to open up a whole new realm of possibilities, and now the entire industry was chasing the technology as if it was the answer to everything, even though its future in areas other than speech and image recognition and machine translation remained unclear. Lu never did persuade the Microsoft brain trust to build a self-driving car, but as the fever enveloped the industry, he convinced them they should at least do something.

The most important figures of the deep learning revolution were already working for the competition. Google had Hinton, Sutskever, and Krizhevsky as well as Hassabis, Legg, and Silver. Facebook had LeCun. Baidu had Andrew Ng. But in a world where a figure like Hinton or Hassabis was an invaluable commodity—a way of understanding the changes ahead, building new technology, attracting top talent, and, on top of everything else, promoting a corporate brand—Microsoft did not have one of its own.

For Qi Lu, the only remaining option was Yoshua Bengio, the third founding father of the deep learning movement who had nurtured a lab at the University of Montreal as Hinton and LeCun toiled in Toronto and New York. Unlike Hinton and LeCun, Bengio specialized in natural language understanding—systems that aimed to master the natural way we humans put words together. He and his students were at the heart of the next big breakthrough, creating a new breed of machine translation alongside Google and Baidu. The rub was that, like LeCun, for whom he had once worked at Bell Labs, he very much believed in academic freedom. By the summer of 2016, he had already rebuffed advances from all the big American tech companies. But Lu believed he could still bring him to Microsoft—and Microsoft was willing to foot the bill. One morning that fall, with the blessing of the company's new chief executive, Satya Nadella, Lu boarded a plane with Li Deng and another Microsoft researcher and flew to Montreal.

They met Bengio in his office at the university, a tiny book-filled room that could barely hold the four of them. Bengio immediately told them he would not join Microsoft, no matter how much money they offered. A man with thick eyebrows and a head of tightly curled salt-and-pepper hair who spoke English with only the hint of a French accent, he carried a seriousness that was both charming and a little intimidating. He said he preferred life in Montreal, where he could speak his native French, and he preferred the openness of academic research, which was still unmatched in the corporate world. But the four of them kept talking. Alongside his work at the university, Bengio was backing other ventures, and when he said he was spending a portion of his time advising a new Canadian start-up called Maluuba, which specialized in conversational systems, this gave Lu another way in. If Microsoft acquired Maluuba, he said, Bengio could spend the same portion of his time advising Microsoft. Before the morning was out, after an email exchange with Nadella, he'd verbally offered to buy the start-up, and Nadella said that if they agreed to a sale, he would fly Bengio and the Maluuba founders to Seattle that evening for a sit-down.

The two founders joined them for lunch at a university café, but they did not fly to Seattle. They turned the offer down, saying their start-up, founded just months earlier, still needed room to grow. Lu continued to press, but they wouldn't budge, and neither would Bengio. He didn't want to talk business. He wanted to talk AI. At one point, as they discussed artificial intelligence and robotics and where all this was headed, he said that the robots of the future would need to sleep. They would need to sleep, he argued, because they would need to dream. His point was that the future of AI research lay with systems that could not only recognize pictures and spoken words but also *generate them on their own.* Dreaming is a vital part of the way humans learn. At night, we replay what we have experienced during the day, burning memories into our brains. The same would one day be true of robots.

As their lunch ended, Lu told them the offer would still be there when they wanted it. He then hobbled out of the café on his cane. Maluuba did join the company about a year later, with Bengio serving Microsoft as a conspicuous consultant. But by then, Lu had left the company. The first round of surgery on his hip hadn't been completely successful: It had left his spine out of alignment, causing pain throughout his body. When he returned from Montreal, and the doctors told him he needed another surgery, he told Nadella that it no longer made sense for him to stay at Microsoft. The recovery would take too long and he couldn't dedicate the time to the company he needed to. Microsoft announced his departure in September 2016. Five months later, he returned to China and joined Baidu as chief operating officer.

PART THREE

TURMOIL

13

DECEIT

"OH, YOU CAN ACTUALLY MAKE PHOTO-REALISTIC FACES."

Ian Goodfellow had interviewed at Facebook in the fall of 2013, listening to Mark Zuckerberg philosophize about DeepMind as they strolled across the campus courtyard. Then he'd turned Zuckerberg down in favor of a job with Google Brain. But at the moment, his professional life was on hold. He'd decided to stay in Montreal for the time being. He was still waiting for his PhD thesis panel to convene, after making the mistake of asking Yann LeCun to join the panel just before Facebook unveiled its new AI lab, and he wanted to see how his relationship would develop with the woman he'd only just started dating. He was also writing a textbook on deep learning, but it wasn't going too well. He mostly sat around drawing baby elephants and posting them to the Internet.

That sense of drift came to an end when one of his university labmates landed a job with DeepMind, and the lab's researchers arranged a goodbye party at a bar just down L'Avenue du Mont-Royal. The bar was called Les 3 Brasseurs, the 3 Brewers. It was the kind of place where twenty people could show up unannounced, push several tables together, sit down, and start drinking large amounts of craft beer. Goodfellow was already buzzed when these researchers started arguing about the best way to build a machine that could create its own photo-realistic images—pictures of dogs or frogs or faces that looked completely real but didn't actually exist. A few labmates were trying to build one. They knew they could train a neural network to recognize images and then flip it upside down so it could *generate* them, too. That's what the DeepMind researcher Alex Graves had done in building a system that could write in longhand. But this worked only so well with detailed, photolike imagery. The results were invariably unconvincing.

Goodfellow's labmates, however, had a plan. They would statistically analyze each image that emerged from their neural network—identify the frequency of certain pixels and their brightness and the way they correlated with other pixels. Then they could compare these stats to what they found in real photos, and that would show their neural network where it went wrong. The rub was that they had no idea how to code all this into their system—it could require billions of statistics. Goodfellow told them the problem was insuperable. "There are just too many different statistics to track," he said. "This is not a programming problem. It's an algorithm design problem."

He offered a radically different solution. What they should do, he explained, was build a neural network that learned *from another neural network.* The first neural network would create an image and try to fool the second into thinking it was real. The second would pinpoint where the first went wrong. The first would try again. And so on. If these dueling neural networks dueled long enough, he said, they could build an image that looked like the real thing. Goodfellow's colleagues were unimpressed. His idea, they said, was even worse than theirs. And if he hadn't been slightly drunk, Goodfellow would have come to the same conclusion. "It's hard enough to train one neural net," a sober Goodfellow would have said. "You can't train one neural net inside the learning algorithm of another." But in the moment, he was convinced it would work.

When he walked into his one-room apartment later that night, his girlfriend was already in bed. She woke up, said hello, and then went back to sleep. So he sat down at a desk next to the bed, in the dark, still slightly drunk and with his laptop screen shining on his face. "My friends are wrong!" he kept telling himself as he pieced his dueling networks together using old code from other projects and training this new contraption on several hundred photos while his girlfriend slept beside him. A few hours later, it was working as he had predicted it would. The images were tiny, no bigger than a thumbnail. And they were a bit blurry. But they looked like photos. It was, he later said, a stroke of luck. "If it hadn't worked, I might have given up on the idea." In the paper he published on the idea, he called them "generative adversarial networks," or GANs. Across the worldwide community of AI researchers, he became "the GANfather."

By the time he joined Google in the summer of 2014, he was promoting GANs as a way of accelerating the progress of artificial intelligence. In describing the idea, he often pointed to Richard Feynman. The blackboard in Feyman's classroom once read: "What I cannot create, I do not understand." This was what Yoshua Bengio, Goodfellow's advisor at the University of Montreal, had argued in a café near the university when wooed by the traveling contingent from Microsoft. Like Hinton, Bengio and Goodfellow believed Feyman's adage applied to machines as well as people: *What artificial intelligence cannot create, it cannot understand.* Creating, they all argued, would help machines understand the world around them. "If an AI can imagine the world in realistic detail—learn how to imagine realistic images and realistic sounds—this encourages the AI to learn about the structure of the world that actually exists," Goodfellow says. "It can help the AI understand the images that it sees or sounds that it hears." Like speech recognition and image recognition and translation, GANs were another leap forward for deep learning. Or, at least, that was what deep learning researchers believed.

In a speech at Carnegie Mellon University in November 2016, Yann LeCun called GANs "the coolest idea in deep learning in the last twenty years." When Geoff Hinton heard this, he pretended to count backwards through the years, as if to make sure GANs weren't any cooler than backpropagation, before acknowledging that LeCun's claim wasn't far from the truth. Goodfellow's work sparked a long line of projects that refined and expanded and challenged his big idea. Researchers at the University of Wyoming built a system that generated tiny but perfect images of insects, churches, volcanos, restaurants, canyons, and banquet halls. A team at Nvidia built a neural network that could ingest a photo of a summer day and turn it into the dead of winter. A group at the University of California–Berkeley designed a system that converted horses into zebras and Monets into van Goghs. They were among the most eye-catching and intriguing projects across both industry and academia. Then the world changed.

IN November 2016—the month Yann LeCun gave a speech calling GANs the coolest idea in deep learning of the last twenty years—Donald Trump defeated Hillary Clinton. And as American life and international politics underwent a seismic shift, so did AI. Almost immediately, the administration's clampdown on immigration stoked concerns over the movement of talent. As the number of international students studying in the United States, already on the decline, now fell sharply, an American science and mathematics community that relied so heavily on foreign talent began to suffer. "We are shooting ourselves in the head," said Oren Etzioni, the chief executive of the Allen Institute for Artificial Intelligence, an influential lab based in Seattle. "Not in the foot. In the head."

The big companies were already expanding their operations abroad. Facebook opened AI labs in both Montreal and Paris, Yann LeCun's hometown. Microsoft ended up buying Maluuba, which became its own lab in Montreal (with Yoshua Bengio as a high-priced advisor). And rather than spend his time in Mountain View, Geoff Hinton opened a Google lab in Toronto. He did this in part so he could care for his wife, who continued to battle cancer. She'd often made the trip to Northern California, where they would spend the weekends in Big Sur, one of her favorite spots. But as her health deteriorated, she needed to be at home. She was adamant that Hinton continue his work, and as he did, a larger ecosystem flourished around him.

The risks of the Trump administration's immigration policy came sharply into focus in April 2017, just three months after he took office, when Hinton helped open the Vector Institute for Artificial Intelligence, a Toronto research incubator. It was backed by $130 million in funding, including dollars from American giants like Google and Nvidia, but it was designed to generate new Canadian start-ups. Prime Minister Justin Trudeau promised $93 million in support of AI research centers in Toronto and Montreal, as well as Edmonton. The career path of one of Hinton's key collaborators, a young researcher named Sara Sabour, exemplified the international nature of AI and how susceptible it was to political interference. In 2013, after completing a computer science degree at the Sharif University of Technology in Iran, Sabour had applied to the University of Washington, hoping to study computer vision and other forms of artificial intelligence, and she had been accepted. But then the U.S. government denied her a visa, apparently because she had grown up and studied in Iran and aimed to specialize in an area, computer vision, that potentially played into military and security technologies. The following year, she enrolled at the University of Toronto, before finding her way to Hinton and Google.

Meanwhile, the Trump administration continued to focus on keeping people out of the country. "Right now, the benefit is captured by American companies," said Adam Segal, a specialist in emerging technologies and national security at the Council on Foreign Relations. "But in the long term, the tech and job creation is not happening in America." Andrew Moore, the dean of computer science at Carnegie Mellon, one of the hubs of AI research in the United States, said the situation was close to keeping him up at night. One of Moore's professors, Garth Gibson, left Carnegie Mellon to take the reins of Toronto's Vector Institute. Seven other professors left for academic posts in Switzerland, where governments and universities were offering far more for this kind of research than they were in the U.S.

But the talent shift wasn't the biggest change wrought by Trump's arrival in the Oval Office. From the moment the election ended, the national media started to question the role of online misinformation in the outcome, raising deep concerns over the power of "fake news." Mark Zuckerberg initially dismissed these concerns during a public appearance in Silicon Valley days after the election, blithely saying it was a "pretty crazy idea" that voters were swayed by fake news. A chorus of reporters, lawmakers, pundits, and private citizens, however, pushed back. The truth was that the problem had been rampant during the election, particularly on Facebook's social network, where hundreds of thousands of people, perhaps even millions, had shared hoax stories with headlines like "FBI Agent Suspected in Hillary Email Leaks Found Dead of Apparent Murder-Suicide" and "Pope Francis Shocks World, Endorses Donald Trump for President." After Facebook revealed that a Russian company with links to the Kremlin had purchased more than $100,000 in ads on the site from four hundred and seventy fake accounts and pages, spreading divisive messages related to race, gun control, gay rights, and immigration, these concerns continued to grow. As they did, they cast GANs and related technologies in a new light. These technologies, it seemed, were a way of generating fake news.

Researchers played right into the narrative. A team from the University of Washington, including one researcher who soon moved to Facebook, used a neural network to build a video that put new words in the mouth of Barack Obama. At a start-up in China, engineers used similar techniques to turn Donald Trump into a Chinese speaker. It was not that fake images were a new thing. People had been using technology to doctor photographs since the dawn of photographs, and in the computer age, tools like Photoshop gave nearly anyone the power to edit both photos and videos. But because the new deep learning methods could learn the task on their own—or at least part of the task—they threatened to make the editing that much easier. Rather than paying farms of people to create and distribute fake images and fake video, political campaigns, nation-states, activists, and insurrectionists could potentially build systems that did the job automatically.

At the time of the election, the full potential of image-manipulating AI was still some months away. As it stood, GANs could generate only thumbnails, and the systems that put words into the mouths of politicians still required a rare expertise, not to mention the added elbow grease. But then, on the first anniversary of Trump's win, a team of researchers at a Nvidia lab in Finland unveiled a new breed of GAN. Called "Progressive GANs," these dueling neural networks could generate full-sized images of plants, horses, buses, and bicycles that seemed like the real thing. But it was the faces that grabbed the attention. After analyzing thousands of celebrity photos, Nvidia's system could generate a face that looked like a celebrity even though it wasn't—a face that reminded you of Jennifer Aniston or Selena Gomez, though it was neither. These invented faces seemed like real faces, with their wrinkles and their pores and their shadows and their very own character. "The technology advanced so quickly," says Phillip Isola, an MIT professor who helped develop these kinds of techniques, "it went from 'O.K., this is a really interesting academic problem, but you can't possibly use this to make fake news. It's just going to produce a little blurry object,' to 'Oh, you can actually make photo-realistic faces.'"

Days after Nvidia unveiled the technology, in the minutes before Ian Goodfellow gave a speech at a small conference in Boston, a reporter asked what it all meant. He acknowledged that anyone could already build fake images with Photoshop, but he also said the task was getting easier. "We're speeding up things that are already possible," he said. Dressed in a black shirt and blue jeans, with a goatee on the end of his chin, his hair combed forward onto his forehead, he somehow looked and spoke like both the nerdiest and the coolest person in the room. As these methods improved, he explained, they would end the era where images were proof that something had happened. "It's been a little bit of a fluke, historically, that we're able to rely on videos as evidence that something really happened," he said. "We used to actually have to think through a story about who said what and who has the incentive to say what, who has credibility, on which issue. And it seems like we're headed back towards those kinds of times." But that would be a hard transition to make. "Unfortunately, people these days are not very good at critical thinking. And people tend to have a very tribalistic idea of who's credible and not credible." There would, at the very least, be a period of adjustment. "There's a lot of other areas where AI is opening doors that we've never opened before. And we don't really know what's on the other side," he said. "In this case, it's more like AI is closing some of the doors that our generation has been used to having open."

This period of adjustment began almost immediately, as someone calling themselves "Deepfakes" started splicing celebrity faces into porn videos and posting them to the Internet. After this anonymous prankster distributed a software app that did the trick, these videos turned up, en masse, across discussion boards and social networks and video sites like YouTube. One used the face of Michelle Obama. Several pulled the trick with Nicolas Cage. Services like Pornhub and Reddit and Twitter soon banned the practice, but not before the idea spilled into the mainstream media. "Deepfake" entered the lexicon, a name for any video doctored with artificial intelligence and spread online.

Even as he helped push AI forward, Ian Goodfellow came to

share the growing concern over its rapid rise, a more immediate concern than the warnings of superintelligence tossed into the zeitgeist by Elon Musk. GANs were only part of it. When Goodfellow first arrived at Google, he began to explore a separate technique called "adversarial attacks," showing that a neural network could be fooled into seeing or hearing things that weren't really there. Just by changing a few pixels in a photo of an elephant—a change imperceptible to the human eye—he could fool a neural network into thinking this elephant was a car. A neural network learned from such a wide array of examples that small and unexpected flaws could creep into its training without anyone ever knowing. The phenomenon was particularly worrisome when you considered that these algorithms were moving into self-driving cars, helping them recognize pedestrians, vehicles, street signs, and other objects on the road. Soon a team of researchers showed that by slapping a few Post-it notes on a stop sign, they could fool a car into thinking it wasn't there. Goodfellow warned that the same phenomenon could undermine any number of other applications. A financial firm, he said, could apply the idea to trading systems, making a few trades designed to fool competitors into dumping a stock—just so the firm could buy shares at a much lower price.

In the spring of 2016, after less than two years at Google, Goodfellow left the company and took this research to the new OpenAI lab, drawn to its stated mission of building ethically sound artificial intelligence and sharing it with the world at large. His work, including both GANs and adversarial attacks, was a natural fit. His aim was to show the effects of these phenomena and how the world should face them. Plus, as the lab's tax filings later showed, he was paid $800,000 for just the last nine months of the year (including a $600,000 signing bonus). But he wasn't at OpenAI much longer than those nine months. The following year, he returned to Google, as Jeff Dean created a new group inside Google Brain dedicated to AI security. Given Goodfellow's high profile across the research community and the wider tech industry, the move was a black eye for OpenAI. It also showed that concerns over the rise of AI were much larger than just one lab.

14

HUBRIS

"I KNEW WHEN I GAVE THE SPEECH THAT THE CHINESE WERE COMING."

In the spring of 2017, a year after the match in Korea, AlphaGo played its next match in Wuzhen, China, an ancient water town eighty miles south of Shanghai along the Yangtze River. With its lily ponds, stone bridges, and narrow boat canals that snaked between rows of small wooden houses topped by rock-tile roofs, Wuzhen is a village meant to look as it had centuries before. Then a two-hundred-thousand-square-foot conference center rises up among the rice paddies. It looks a lot like the wooden houses spread across the village, except it's the size of a soccer stadium. Its roof spans more than 2.5 trillion tiles. Built to host the World Internet Conference, an annual gathering where Chinese authorities trumpeted the rise of new Internet technologies and marked the ways they would regulate and control the spread of information, it was now hosting a match between AlphaGo and the Chinese grandmaster Ke Jie, the current number-one-ranked Go player in the world.

The morning of the first game, inside a private room down a side hall from the cavernous auditorium where the match was due to be played, Demis Hassabis sat in a plush, oversized, cream-colored chair in front of a wall painted like an afternoon sky. This was the theme across the building: cloud-strewn afternoon skies. Wearing a dark-blue suit with a small round royal-blue pin on the lapel and no necktie—suddenly looking older and more polished than he had the year before—Hassabis said AlphaGo was now much more talented. Since the match in Korea, DeepMind had spent months improving the machine's design, and AlphaGo had spent many more playing game after game against itself, learning entirely new skills through digital trial and error. Hassabis was confident the machine was now immune to the kind of sudden meltdown that arrived during the fourth game in Korea, when Lee Sedol, with Move 78, exposed a gap in its knowledge of the game. "A big part of what we were trying to do with the new architecture was close the knowledge gap," Hassabis said. The new architecture was also more efficient. It could train itself in a fraction of the time, and once trained, it could run on a single computer chip (a Google TPU, naturally). Though Hassabis didn't exactly say so, it was clear, even then, before the first move of the first game, that the nineteen-year-old Ke Jie had no chance of winning. The Google powers that be had arranged the match as a valedictory lap for AlphaGo—and as a way of easing the company's return to China.

In 2010, Google had suddenly and dramatically pulled out of China, moving its Chinese language search engine to Hong Kong after accusing the government of hacking into its corporate network and eavesdropping on the Gmail accounts of human-rights activists. The message, from Larry Page, Sergey Brin, and Eric Schmidt, had been that Google was leaving not only because of the hack but because it could no longer abide government censorship of news stories, websites, and social media on the Google search engine. The new Hong Kong servers sat outside what Westerners called "the Great Firewall of China." Several weeks later, as Google knew they would, officials in Beijing brought the curtain down on these Hong Kong machines, barring access to anyone on the mainland, and this was the way it stayed for seven years. But by 2017, Google was not what it once had been. In creating Alphabet, the umbrella company that oversaw Google, DeepMind, and various sister companies, Page and Brin had stepped away from the tech giant they had built over the previous two decades, putting each company in the hands of other executives and, it seemed, inching toward an early retirement. Under its new CEO, Sundar Pichai, Google was having second thoughts about China. The market was too big to ignore. There were more people on the Internet in China than people living in the United States—about 680 million—and that number was growing at a rate no other country could match. Google wanted back in.

The company saw AlphaGo as the ideal vehicle. In China, Go was a national game. An estimated 60 million Chinese had watched the match against Lee Sedol as it streamed over the Internet. And as Google eyed the Chinese market, one of its primary aims was to promote its expertise in artificial intelligence. Even before the AlphaGo match with Lee Sedol, executives at Google and DeepMind had discussed the possibility of a second match in China that would pave a path back into the market for the Google search engine and the company's many other online services, and after all the ballyhoo in Korea, the idea snowballed. They saw it as "Ping-Pong diplomacy," akin to the United States staging table tennis matches in China during the 1970s as a way of easing diplomatic relations between the two countries. Google spent the next year planning AlphaGo's visit to China, meeting with the national minister of sport and arranging for multiple Internet and television services to broadcast the match. Sundar Pichai took three trips to China before the match, personally meeting with Ke Jie. The two posed for a photo on the Great Wall. Alongside the match itself, Google organized its own artificial intelligence symposium in the same conference center, after the first game of the match and before the second. Both Jeff Dean and Eric Schmidt made the trip to China, and both were set to give speeches during this one-day mini-conference. Dozens of Chinese journalists descended on Wuzhen for the match, and many more traveled from around the world. When Demis Hassabis walked through the conference center before the first game, they photographed him like he was a pop star.

Later that morning, as Hassabis described the ongoing evolution of AlphaGo in the room with the afternoon sky painted on the wall, Ke Jie made the game's first move in the auditorium, a few hundred feet away. Across the country, millions were poised to watch from afar. But they couldn't. With a private order sent to all Chinese media in Wuzhen, the government shut down all Internet and television broadcasts of the match. In China, the match would be seen only by the few hundred people allowed through the security gates leading into Wuzhen and then past the armed guards, electronic badge readers, and metal detectors at the front of the conference center. Newspapers and news sites could cover the match, but if they did, they were forbidden to use the word "Google." As the first few moves of the match played out, Hassabis continued to describe the future of Google and DeepMind and its technologies. He did not mention the blackout.

CHINA was hardly new to deep learning. In early December 2009, Li Deng had once again made the drive from the NIPS conference in Vancouver to the NIPS workshops in Whistler. A year after running into Geoff Hinton inside the Whistler Hilton and stumbling onto his research with deep learning and speech recognition, Deng had organized a new workshop around the idea at the same spot high in the Canadian mountains. He and Hinton would spend the next few days explaining the finer points of "neural speech recognition" to the other researchers gathered in Whistler, walking them through the prototype under way at the Microsoft lab in Redmond. As he drove north, winding through the mountain roads, Deng carried three of these researchers in his SUV. One of them was Kai Yu, the man who would later talk the Baidu brain trust into bidding big money for Geoff Hinton.

Like Deng, Yu was born and educated in China before taking a research job in the United States. Deng worked for Microsoft outside Seattle, while Yu was part of a Silicon Valley lab inside the hardware maker NEC, but they were both part of the small community of researchers who turned up at academic gatherings like the one in Whistler. That year, they were also in the same carpool. Yu already knew Geoff Hinton, from having organized a deep learning workshop alongside Yann LeCun and Yoshua Bengio the previous summer in Montreal, and now, as the SUV climbed the mountains, he was headed for a larger gathering dedicated to the same idea. As deep learning emerged from academia and moved into industry, Kai Yu was as close to the moment as anyone, and when he returned to China the following year, he brought the idea with him.

As Deng and Hinton and Hinton's students remade speech recognition at Microsoft and IBM and Google, Yu did the same at

Baidu. Within months, his work caught the eye of Robin Li, the company's chief executive, who boasted of the power of the technology in an email message broadcast across the company. This was why Baidu was willing to join the chase for Hinton and his students in 2012, bidding tens of millions of dollars during the auction in Lake Tahoe—and why Kai Yu was so optimistic that, even though Baidu lost out on that occasion, it would stay in the race for deep learning.

He was not the only one whispering in the ear of Robin Li, the Baidu chief executive. Li was also an old friend of Qi Lu, the Microsoft executive who would later break his hip riding a backwards bicycle. They'd known each other for more than two decades. Each year, alongside several other Chinese and Chinese-American executives, including the CEO of the Beijing computer giant Lenovo, they would meet for a kind of cross-border summit at the Ritz-Carlton hotel in Half Moon Bay, California, just down the coast from San Francisco, spending several days discussing the latest shifts in the technological landscape. At their meeting in the wake of the auction for Hinton and his students, deep learning was the topic of conversation. Inside the Ritz-Carlton, a resort that looked across the Pacific Ocean, Qi Lu mapped out a convolutional neural network on a whiteboard, telling Li and the others that "CNN" now meant something very different. That same year, Baidu opened its first outpost in Silicon Valley, not far from Half Moon Bay, hoping to attract North American talent. It was called the Institute of Deep Learning. Kai Yu told a reporter its aim was to simulate "the functionality, the power, and the intelligence" of the human brain. "We are making progress day by day," he said.

The following spring, a short drive from this new lab, Yu sat down for breakfast at the Palo Alto Sheraton with Andrew Ng. They met again for dinner that evening, and Ng signed on with

Baidu after flying to China for a meeting with Robin Li, the CEO. The man who had founded Google's deep learning lab was now running much the same enterprise at one of China's biggest companies, overseeing labs in both Silicon Valley and Beijing. By the time Google descended on Wuzhen in the spring of 2017 for the Go match designed to ease its return to the Chinese market, Yu and Ng and their researchers had already pushed deep learning into the heart of the Baidu empire, using it, like Google, to choose search results and target ads and translate between languages. After hiring a key engineer from chip maker Nvidia, Baidu had built its own giant cluster of GPUs. And Yu had already left the company to launch a Chinese start-up that aimed to build a new kind of deep learning chip along the lines of the Google TPU.

When Google chairman Eric Schmidt took the stage in Wuzhen, between games one and two of the AlphaGo match, he acted as if none of this had ever happened. Sitting in a chair next to his Chinese interviewer, one leg crossed over the other, a tiny device wrapped around his ear feeding him an English translation of each question, Schmidt said the world was entering "the age of intelligence," meaning, of course, artificial intelligence. Using a new software creation called TensorFlow, he said, Google had built AI that could recognize objects in photos and identify spoken words and translate between languages. Addressing his audience as he always did—as if he knew more than anyone else in the room, about both the past and the future—he described this software as the biggest technological change of his lifetime. He boasted that it could reinvent China's biggest Internet companies, including Alibaba, Tencent, and Baidu, claiming it could target their online ads, predict what their customers wanted to buy, and decide who should get a line of credit. "All of them would be better off," Schmidt said, "if they used TensorFlow."

Conceived and designed by Jeff Dean and his team, TensorFlow was the successor to DistBelief, the sweeping software system that trained deep neural networks across Google's global network of data centers. But that was not all. After deploying the software in its own data centers, Google had *open-sourced* this creation, freely sharing the code with the world at large. This was a way of exerting its power across the tech landscape. If other companies, universities, government agencies, and individuals used Google's software as they, too, pushed into deep learning, their efforts would feed the progress of Google's own work, accelerating AI research across the globe and allowing the company to build on this research. It would breed a whole new community of researchers and engineers Google could hire from. And it would bootstrap the business that Google viewed as its future: cloud computing.

When Eric Schmidt delivered his message from the stage in Wuzhen, more than 90 percent of Google's revenues still came from online advertising. But Google had looked into the future and realized that as cloud computing became a more reliable and lucrative alternative, offering anyone off-site computing power and data storage, the company was in an ideal position to exploit its commercial potential. Google had harnessed enormous amounts of computing power inside its data centers, and selling access to this power could be prodigiously profitable. At the moment, this rapidly growing market was dominated by Amazon, whose cloud revenue would top $17.45 billion in 2017. But TensorFlow offered the hope that Google could take on its rival tech giant. If TensorFlow became the de facto standard for building artificial intelligence, Google believed, it could lure the market onto its cloud computing services. Google's data center network would be, in theory, the most efficient means of running TensorFlow, in part because the company offered a chip built just for deep learning. As Schmidt expounded on the virtues of TensorFlow and urged the giants of the Chinese tech industry to embrace it, Google had already built a second incarnation of its TPU chip, one designed to *both train neural networks and run them once they were trained*. It was also working to build a new AI lab in Beijing that, the company hoped, would help push China toward TensorFlow and the new chip and, ultimately, the Google cloud. The lab would be led by a new Google hire named Fei-Fei Li, who had been born in Beijing before emigrating to the United States as a teenager. As she said: "There was a growing community of AI research talent in China. The lab might allow us both to tap into that pool and to spread the use of TensorFlow (and Google Cloud) more broadly in the country."

Sitting onstage in Wuzhen, telling the auditorium that TensorFlow could reinvent the top Chinese companies, Eric Schmidt didn't mention the AI lab or even the Google cloud. But his message was clear: Alibaba and Tencent and Baidu would be better off if they used TensorFlow. What he didn't say was that he knew Google would benefit far more. What he didn't realize was that his message to the Chinese was hopelessly naïve.

China's tech giants had already embraced deep learning. Andrew Ng had been building labs at Baidu for years, and, like Google, he was erecting a vast network of specialized machines to feed new experiments. Similar work was brewing at Tencent. In any case, even if it did need Google's help, China was unwilling to take it. The government, after all, had blacked out the match in Wuzhen. It did not take long for Schmidt to realize just how naïve his message had been. "I knew when I gave the speech that the Chinese were coming. I did not understand at the time how totally effective some of their programs would be," he says. "I honestly just didn't understand. I think most Americans wouldn't understand. I'm not going to misunderstand in the future."

The week in Wuzhen was not what anyone at Google had envisioned. On the morning of the first game with Ke Jie, sitting in front of that painted afternoon sky, Demis Hassabis said AlphaGo would soon grow even more powerful. His researchers were building a version that could master the game *entirely* on its own. Unlike the original incarnation of AlphaGo, it didn't need to learn its initial skills by analyzing moves from professional players. "It removes more and more of the human knowledge," Hassabis says. Learning solely from trial and error, it could master not only Go but other games, like chess and Shōgi, another ancient Eastern strategy game. With this kind of system—a more general form of AI that could learn myriad tasks on its own—DeepMind could transform an increasingly wide and varied array of technologies and industries. As Hassabis put it, the lab could help manage resources inside data centers and power grids and accelerate scientific research. The message, once again, was that DeepMind's technology would boost human performance. This, he said, would be obvious in the match with Ke Jie. Like the rest of the world's top Go players, the Chinese grandmaster was now mimicking the style and the skills of AlphaGo. His play was improving because he was learning from the machine.

From the opening move, Ke Jie did indeed play like AlphaGo, opening with the same "3–3 point" gambit the machine had introduced to the game. But the result was never in doubt. Wearing a dark suit with a bright blue tie and black-rimmed glasses, the nineteen-year-old Jie had a habit of playing with the hair on his head as he thought through each new move, pinching short strands between his thumb and index finger and then twirling them around one and then the other. Sitting onstage in the Wuzhen auditorium, he twirled his hair for more than twelve hours over three different days. After losing the first game, he said AlphaGo was "like a God of a Go player." Then he played out two more, losing those as well.

When Lee Sedol lost in Korea, it was a celebration of both AI and humanity. When Ke Jie lost, it underlined what those at the top rungs of the Chinese government did not want underlined—that the West was sprinting ahead in the race to the future. AlphaGo did not just win handily. It won against a Chinese grandmaster in China. And between games one and two, Eric Schmidt spent thirty minutes patronizing the country and its biggest Internet companies.

Two months later, the Chinese State Council unveiled its plan to become the world leader in artificial intelligence by 2030, aiming to surpass all rivals, including the United States, as it built a domestic industry worth more than $150 billion. China was treating artificial intelligence like its own Apollo program. The government was preparing to invest in moonshot projects across industry, academia, and the military. As two university professors who were working on the plan told the *New York Times*, AlphaGo versus Lee Sedol was China's Sputnik moment.

China's plan echoed a blueprint the Obama administration laid down just before leaving office, often using much of the same language. One difference was that the Chinese government was already investing big money in this effort. One municipality had promised $6 billion. The other difference was that the Chinese plan hadn't been discarded by a new government, as the Obama plan was abandoned by the Trump administration. China was working to coordinate government, academia, and industry in a monolithic push toward AI, while the new administration in the U.S. was leaving the push to industry. Google was the world leader in the field and other American companies were not far behind, but it was unclear what that meant for the United States as a whole. So much of the AI talent, after all, had moved into industry, leaving academia and government behind. "The worry for the U.S. is that

China will be putting more money into research than they are," Geoff Hinton said. "The U.S. is cutting the money for basic research, which is like eating the seed corn."

What was clear was that Google would not make much headway in China. Later that year, at an event in Shanghai, Fei-Fei Li unveiled what she called the Google AI China Center, and the company continued to push TensorFlow, sending engineers to private events where they taught industry and university researchers to use the software. But it would need government approval to launch new Internet services in China. And China already had its own search engine, its own cloud computing services, its own AI labs, even its own TensorFlow. Called PaddlePaddle, it was built by Baidu.

With his speech in Wuzhen, Eric Schmidt had underestimated the Chinese. The big Chinese tech companies—and the country as a whole—were much further along and carried far more potential than he realized. He was wrong to so blithely tell the country it needed Google and TensorFlow. But now, he realized, the spread of this technological platform—a foundational way of building and running any artificial intelligence service—was more important than ever. It was vitally important not only for Google but for the United States and its escalating economic battle with China. "One of the things that is not understood at all by the conventional forces in society, if you will, is that America benefits enormously from these global platforms—global platforms that are built in America, whether it's the Internet itself, email, Android, the iPhone, et cetera," Schmidt says. If a company, or indeed a country, controlled the platform, it controlled what ran on top of it. A creation like Google's TensorFlow was the latest example. "It's a global platform competition, and it is extremely important that the platforms be invented in America. Platforms establish a base by which innovation occurs in the future."

JUST after the match in Wuzhen, Qi Lu joined Baidu. There, he did what he'd wanted to do at Microsoft: build a self-driving car. The company launched its project years after Google, but Lu was sure it would put cars on the road far faster than its American rival. This wasn't because Baidu had better engineers or better technology. It was because Baidu was building its car in China. In China, government was closer to industry. As Baidu's chief operating officer, he was working with five Chinese municipalities to remake their cities so that they could accommodate the company's self-driving cars. "There is no question in my mind this will get commercialized way sooner in China than in the United States. The government sees this as a chance for China's auto industry to leapfrog," he told a reporter during one of his regular trips back to the U.S. "Spurring investment is one thing. Actively working with the companies to craft policy regimes is another." At the moment, he explained in his axiomatic English, street signs were the infrastructure that allowed the sensors of a car to navigate the road, and the sensors of a car were human eyeballs. But all this would change, and it would change much quicker in China. In the future, he said, the sensors of a car would be laser-based lidar sensors, radar, and cameras, and there would be new kinds of street signs designed just for these sensors.

China's other advantage, he said, was data. In each socioeconomic era, he liked to say, there was one primary means of production. In the agricultural era, it was about the land. "It doesn't matter how many people you have. It doesn't matter how brilliant you are. You cannot produce more if you do not have more land." In the industrial era, it was about labor and equipment. In the new era, it was about data. "Without data, you cannot build a speech recognizer. It does not matter how many people you have. You may have one million brilliant engineers, but you won't be able to build a system that understands language and can have a dialogue. You won't be able to build a system that can recognize images like I do right now." China would rule this era because it would have more data. Because its population was bigger, it would produce more data, and because its attitudes toward privacy were so different, it would have the freedom to pool this data together. "People are less privacy sensitive. The need for privacy is universal, but the treatment is very different in China, the policy regime is different."

Even if the big Chinese companies and universities were technologically behind their American rivals at the moment—and that was debatable—the gap didn't matter all that much. Thanks to the influence of academics like Geoff Hinton and Yann LeCun in the West, the big American companies were openly publishing most of their important ideas and methods and even sharing software. Their ideas, methods, and software were available to anyone, including anyone in China. Ultimately, the primary difference between East and West was the data.

For Qi Lu, all this meant that China would be the first not just to produce driverless cars but also to find a cure for cancer. That, too, he believed, was a product of data. "There is no doubt in my mind," he said.

15

BIGOTRY

"GOOGLE PHOTOS, Y'ALL FUCKED UP. MY FRIEND IS NOT A GORILLA."

One Sunday in June 2015, Jacky Alciné was sitting in the room he shared with his younger brother, scrolling through a long stream of tweets about the Black Entertainment Television awards. Their apartment in the Crown Heights section of Brooklyn didn't have cable TV, so he couldn't watch the awards, but he could at least read the running Twitter commentary on his laptop computer. As he ate a bowl of rice, a friend sent him an Internet link for some snapshots she'd posted to the new Google Photos service. Alciné, a twenty-two-year-old software engineer, had used the service in the past, but not since Google had unveiled a new version several days earlier. The new Google Photos could analyze your snapshots and automatically sort them into digital folders based on what was pictured in each one. One folder might be "dogs," another "birthday party," a third "beach trip." It was a way of browsing images and instantly searching them, too. If you typed "tombstone," Google could automatically find all photos that included a tombstone. When Alciné clicked on the link and opened the service, he couldn't help but notice that his own photos were already reorganized—and that one of the folders read "gorillas." This made no sense to him, so he opened the folder, and when he did, he found more than eighty photos he'd taken nearly a year earlier of a friend during a concert in nearby Prospect Park. His friend was African-American, and Google had labeled her as a "gorilla."

He might have let it go if Google had mistakenly tagged just one photo. But this was *eighty photos*. He took a screenshot and posted it to Twitter, which he thought of as "the world's largest cafeteria room," a place where anyone could show up and get anyone's attention about anything. "Google Photos, y'all fucked up," he wrote. "My friend is not a gorilla." A Google employee sent him a note almost immediately, asking for access to his account, so the company could understand what went wrong. In the press, Google spent the next several days apologizing, saying it was taking immediate action to ensure this would never happen again. It removed the "gorilla" tag from the service entirely, and that was the status quo for years. Five years later, the service still blocked anyone from searching on the word "gorilla."

The trouble was that Google had trained a neural network to recognize gorillas, feeding it thousands of gorilla photos, without realizing the side effects. Neural networks could learn tasks that engineers could never code into a machine on their own, but in training these systems, the onus was on the engineers to choose the right data. What's more, when the training was done, even if these engineers were careful with their choices, they might not understand everything the machine had learned, just because the training happened on such a large scale, across so much data and so many calculations. As a software engineer himself, Jacky Alciné understood the problem. He compared it to making lasagna. "If you mess up the lasagna ingredients early, the whole thing is ruined," he says. "It is the same thing with AI. You have to be very intentional about what you put into it. Otherwise, it is very difficult to undo."

IN a photo of the Google Brain team taken just after the Cat Paper was published in the summer of 2012, when Geoff Hinton was officially a (sixty-four-year-old) intern at the lab, he and Jeff Dean held a giant digital image of a cat. About a dozen other researchers crowded around them. One was Matt Zeiler, a young man in a black short-sleeved polo shirt and faded blue jeans, with a big smile, a shaggy head of hair, and a few days of stubble on his chin. Zeiler studied in the deep learning lab at NYU before landing his own internship at Google Brain the same summer. A year later, he won the ImageNet contest, following in the footsteps of Hinton, Krizhevsky, and Sutskever. Many hailed him as a rock star in what was now the industry's hottest field. Alan Eustace called to offer a big-money job at Google, but as Zeiler often told reporters, he turned Eustace down to start his own company.

The company was called Clarifai. Based in New York City, in a small office not far from the deep learning lab at NYU, it built technology that could automatically recognize objects in digital images, searching through photos of shoes, dresses, and handbags on a retail website or identifying faces in video footage streaming from a security camera. The idea was to duplicate the image recognition systems the likes of Google and Microsoft had spent the last few years building inside their own AI labs—and then sell it to other businesses, police departments, and government agencies.

In 2017, four years after the company was founded, Deborah

Raji sat at a desk inside the Clarifai offices in lower Manhattan. An unforgiving fluorescent light covered her and her desk and the beer fridge in the corner and all the other twenty-somethings wearing headphones as they stared at oversized computer screens. Raji was staring at a screen filled with faces—images the company used to train its face recognition software. As she scrolled through page after page of these faces, she couldn't help but see the problem. Raji was a twenty-one-year-old black woman from Ottawa. Most of the images—over 80 percent—were of white people. Almost as striking, over 70 percent of those white people were male. When the company trained its system on this data, it might do a decent job of recognizing white people, Raji thought, but it would fail miserably with people of color, and probably women, too.

The problem was endemic. Matt Zeiler and Clarifai were also building what was called a "content moderation system," a tool that could automatically identify and remove pornography from the sea of images people posted to online social networks. The company trained this system on two sets of data: thousands of obscene photos pulled from online porn sites, and thousands of G-rated images purchased from stock photo services. The idea was that their system would learn to tell the difference between the pornographic and the anodyne. The problem was that the G-rated images were dominated by white people, and the pornography were not. As Raji soon realized, the system was learning to identify black people as pornographic. "The data we use to train these systems *matters*," she says. "We can't just blindly pick our sources."

The roots of the problem stretched back years, at least to the moment when someone started choosing images for the stock photo services Clarifai was feeding into its neural networks. It was the same issue that plagued all popular media: It was homogeneous.

Now the risk was that AI researchers using such data would amplify the issue when training automated systems. This much was obvious to Raji. But to the rest of the company, it was not. The people choosing the training data—Matt Zeiler and the engineers he hired at Clarifai—were mostly white men. And because they were mostly white men, they didn't realize their data was biased. Google's gorilla tag should have been a wake-up call for the industry. It was not.

It took other women of color to take this fundamental problem public. Timnit Gebru, who was studying artificial intelligence at Stanford University under Fei-Fei Li, was the Ethiopia-born daughter of an Eritrean couple who had immigrated to the U.S. At NIPS, as she entered the main hall for the first lecture and looked out over the hundreds of people seated in the audience, row after row of faces, she was struck by the fact that while some were East Asian and a few were Indian and a few more were women, the vast majority were white men. Over fifty-five hundred people attended the conference that year. She counted only six black people, all of whom she knew, all of whom were men. And this was not an American or a Canadian conference. It was an international gathering in Barcelona. The problem that Deborah Raji recognized at Clarifai pervaded both industry and academia.

When she returned to Palo Alto, Gebru told her husband what she had seen, and she decided it could not go unremarked upon. Her first night back, sitting cross-legged on the couch with her laptop, she put the conundrum into a Facebook post:

> I'm not worried about machines taking over the world. I'm worried about groupthink, insularity and arrogance in the AI community. Especially with the current hype and demand for people in the field. These things are already causing problems we should worry about right now.

> Machine learning is used to figure out who should get higher interest rates, who is more "likely" to commit a crime and therefore get harsher sentencing, who should be considered a terrorist etc. Some computer vision algorithms we take for granted only work well on people who look a certain way. We don't need to speculate about grand destructions that will happen in the future. AI is working for a certain very small segment of the world population. And the people creating it are from a very minuscule segment of the world population. Certain segments of the population are actively harmed by it. Not only because the algorithms work against them, but also because their jobs are automated. These people are actively excluded from entering a high paying field that is removing them from the workforce. I've heard many talk about diversity as if it is some sort of charity. I see companies and even individuals using it as a PR stunt while paying lip service to it. Because it is the language du jour, "we value diversity" is something you're supposed to say. AI needs to be seen as a system. And the people creating the technology are a big part of the system. If many are actively excluded from its creation, this technology will benefit a few while harming a great many.

This mini-manifesto spread across the community. In the months that followed, Gebru built a new organization called Black in AI. After finishing her PhD, she was hired at Google. The next year, and each year after that, Black in AI ran its own workshop at NIPS. By then, NIPS was no longer called NIPS. After many researchers protested that the name contributed to an environment that was hostile to women, the conference organizers changed the name to NeurIPS.

One of Gebru's academic collaborators was a young computer scientist named Joy Buolamwini. A graduate student at the Massachusetts Institute of Technology in Cambridge who had recently finished a Rhodes Scholarship in England, Buolamwini came from a family of academics. Her grandfather specialized in medicinal chemistry, and so did her father. Born in Edmonton, Alberta, where her father was finishing his PhD, she grew up wherever his research took him, including labs in Africa and across the American South. In the mid-'90s, when she visited his lab while still in elementary school, he mentioned he was dabbling in neural networks for drug discovery—and she had no idea what that meant. After studying robotics and computer vision as an undergraduate, she gravitated toward face recognition, and neural networks resurfaced in a very different way. The literature said that, thanks to deep learning, face recognition was reaching maturity, and yet, when she used it, she knew it wasn't. This became her thesis. "This was not just about facial analysis technology but about the evaluation of facial analysis technology," she says. "How do we establish progress? Who gets to decide what progress means? The major issue that I was seeing was that the standards, the measures by which we decided what progress looked like, could be misleading, and they could be misleading because of a severe lack of representation of who I call the under-sampled majority."

That October, a friend invited her for a night out in Boston with several other women, saying: "We'll do masks." Her friend meant skincare masks at a local spa, but Buolamwini assumed Halloween masks. So she carried a white plastic Halloween mask to her office that morning, and it was still sitting on her desk a few days later as she struggled to finish a project for one of her classes. She was trying to get a face detection system to track her face, and no matter what she did, she couldn't quite get it to work. In her frustration, she picked up the white mask from her desk and pulled it over her head. Before it was all the way on, the system recognized her face—or, at least, it recognized the mask. "*Black Skin, White Masks*," she says, nodding to the 1952 critique of historical racism from the psychiatrist Frantz Fanon. "The metaphor becomes the truth. You have to fit a norm and that norm is not you."

Soon Buolamwini started exploring commercial services designed to analyze faces and identify characteristics like age and gender. This included tools from Microsoft and IBM. As Google and Facebook pushed their face-recognition technologies into smartphone apps, Microsoft and IBM joined Clarifai in offering similar services to businesses and government agencies. Buolamwini found that when these services read photos of lighter-skinned men, they misidentified their gender only about 1 percent of the time. But the darker the skin in the photo, the larger the error rate, and it rose particularly high with images of women with dark skin. Microsoft's error rate was about 21 percent. IBM's was 35.

Published in the winter of 2018, her study drove a larger backlash against face recognition technology and, particularly, its use in law enforcement. The risk was that the technology would falsely identify certain groups as potential criminals. Some researchers argued that the technology could not be properly controlled without government regulation. Soon, the big companies had no choice but to acknowledge the groundswell of opinion. In the wake of the MIT study, Microsoft's chief legal officer said the company had turned down sales to law enforcement when there was concern it could unreasonably infringe on people's rights, and he made a public call for government regulation. That February, Microsoft backed a bill in Washington State that would require notices to be posted in public places using facial recognition tech and ensure that government agencies obtain a court order when looking for specific people. Tellingly, the company did not back other legislation that provided much stronger protections. But attitudes were at least beginning to shift.

While still at Clarifai, after noticing her work on racial and gender bias, Deborah Raji contacted Buolamwini, and they began to collaborate, with Raji eventually moving to MIT. Among other projects, they started testing face technology from a third American tech giant: Amazon. Amazon had moved beyond its online retailer roots to become the dominant player in cloud computing and a major player in deep learning. Just recently, the company had started to market its face technologies to police departments and government agencies under the name Amazon Rekognition, revealing that early customers included the Orlando Police Department in Florida and the Washington County Sheriff's Office in Oregon. Then Buolamwini and Raji published a new study showing that an Amazon face service also had trouble identifying the gender of female and darker-skinned faces. According to the study, the service mistook women for men 19 percent of the time and misidentified darker-skinned women for men 31 percent of the time. For lighter-skinned males, the error rate was zero.

But Amazon responded differently from Microsoft and IBM. It, too, called for government regulation of face recognition. But instead of engaging with Raji and Buolamwini and their study, it attacked them, in private emails and with public blog posts. "The answer to anxieties over new technology is not to run 'tests' inconsistent with how the service is designed to be used, and to amplify the test's [*sic*] false and misleading conclusions through the news media," Amazon executive Matt Wood wrote in a blog post that disputed the study and the *New York Times* article that described it.

This was a product of the ingrained corporate philosophy that drove Amazon. The collective company was insistent that outside voices not disrupt its own beliefs and attitudes. But in dismissing the study, Amazon also dismissed a very real problem. "What I learned is that you don't have to have the truth if you are a trillion-dollar company," Buolamwini says. "You are the bully on the block. What you say is what it is."

BY then, Meg Mitchell had built a team inside Google dedicated to "ethical AI." The researcher who was part of the early deep learning efforts inside Microsoft Research, Mitchell had grabbed the community's attention when she gave an interview to *Bloomberg News* saying that artificial intelligence suffered from a "sea of dudes" problem, estimating she had worked with hundreds of men over the past five years and about ten women. "I do absolutely believe that gender has an effect on the types of questions that we ask," she said. "You're putting yourself in a position of myopia." Mitchell and Timnit Gebru, who joined her at Google, were part of a growing effort to lay down firm ethical frameworks for AI technologies, looking at bias, surveillance, and the rise of automated weapons. Another Googler, Meredith Whittaker, a product manager in the company's cloud computing group, helped launch a research organization at NYU. A consortium of companies, from Google to Facebook to Microsoft, built an organization called the Partnership on AI. Organizations like the Future of Life Institute (founded by Max Tegmark at MIT) and the Future of Humanity Institute (founded by Nick Bostrom at Oxford) were also concerned with the ethics of AI, but they focused on existential threats of the distant future. The new wave of ethicists focused on more immediate matters.

For both Mitchell and Gebru, the bias problem was part of the larger issue across the tech industry. Women struggled to exert their influence in all tech fields, facing extreme bias in the workplace and sometimes harassment. In the field of artificial intelligence, the problem was more pronounced, and potentially more dangerous. All this was why they penned an open letter to Amazon.

In the letter, they refuted the arguments that Matt Wood and Amazon hurled against Buolamwini and Raji. They insisted that the company rethink its approach. And they called its bluff on government regulation. "There are no laws or required standards to ensure that Rekognition is used in a manner that does not infringe on civil liberties," they wrote. "We call on Amazon to stop selling Rekognition to law enforcement." Their letter was signed by twenty-five artificial intelligence researchers across Google, DeepMind, Microsoft, and academia. One of them was Yoshua Bengio. "It was terrifying when it was just us against this big corporation," Raji says. "But it was heartwarming when the community defended our work. I felt it was no longer me and Joy versus Amazon. It was research—hard scientific research—versus Amazon."

16

WEAPONIZATION

"YOU PROBABLY HEARD ELON MUSK AND HIS COMMENT ABOUT AI CAUSING WW3."

Inside the Clarifai offices in lower Manhattan in the fall of 2017, the windows of a room in the far corner were papered over, and a sign on the door read "The Chamber of Secrets." A knowing nod to the second book in the Harry Potter series, the words were written in longhand, and the sign was slightly askew. Behind the door, a team of eight engineers was working on a project they weren't allowed to discuss with the rest of the company. Even the engineers themselves didn't quite understand what they were working on. They knew they were training a system to automatically identify people, vehicles, and buildings in video footage shot somewhere in the desert, but they didn't know how this technology would be used. When they asked, founder and chief executive Matt Zeiler described it as a government project involving "surveillance." He said it would "save lives."

As Clarifai moved into larger offices and engineers dug through the digital files stored on the company's internal computer network, turning up a few documents that mentioned a government contract, their work slowly came into focus. They were building technology for the U.S. Department of Defense as part of something called Project Maven. The idea, it seemed, was to build a system that could identify targets for drone strikes. But the particulars were still unclear. They couldn't tell if the technology would be used to kill or if it would help avoid the killing, as Zeiler said. It wasn't clear if this was a way of automating air strikes or feeding information to human operators before they pulled the trigger on their own.

Then, one afternoon in late 2017, three military personnel in civilian clothes walked into the Clarifai offices and met with a few of the engineers in another room behind closed doors. They wanted to know how precise the technology could be. First, they asked if it could identify a particular building, like a mosque. Mosques, they said, were often converted into military headquarters by terrorists and insurgents. Then they asked if it could differentiate men from women. "What do you mean?" one of the engineers asked. Out in the field, the military personnel explained, it was generally possible to distinguish men (who wore pants) from women (who wore dresses down to the ankle) because of the gap between the legs of the men. They were allowed to shoot the men, they said, not the women. "Sometimes, the men try to fool us by wearing dresses, but that doesn't matter," one of them said. "We are still going to kill all those motherfuckers."

ON Friday, August 11, 2017, Defense Secretary James Mattis sat down at a boardroom table inside Google headquarters in Mountain View. Bouquets of white gardenias ran down the middle of the table, and four pots of coffee sat on a ledge against an emerald-green wall, beside several trays of pastries. The new Google chief executive, Sundar Pichai, sat on the other side of the table. So did Sergey Brin, general counsel Kent Walker, and the head of artificial intelligence, John "J.G." Giannandrea, the man who pushed forty thousand GPU boards into the Google data centers to accelerate the company's AI research. Several others lined the room, including various Defense Department staffers and executives from Google's cloud computing group. Most of the DoD staffers wore suits and ties. Most of the Googlers wore suits and no ties. Sergey Brin wore a white T-shirt.

Mattis was on a West Coast tour, visiting several big tech companies in both Silicon Valley and Seattle as the Pentagon explored its options for Project Maven. Launched by the Defense Department four months earlier, Project Maven was an effort to accelerate the DoD's use of "big data and machine learning." It was also called the Algorithmic Warfare Cross-Functional Team. The project depended on the support of companies like Google that had spent the last several years assembling the expertise and the infrastructure needed to build deep learning systems. This was the way the Defense Department typically built new technology—alongside private industry—but the dynamic was different from in the past. Google and the other companies that controlled the country's AI talent were not traditional military contractors. They were consumer tech companies just beginning to embrace military work. Moreover, now that Donald Trump was in the White House, workers inside these companies were increasingly wary of government projects. Google was particularly sensitive to the tensions here, as its unique corporate culture allowed—or even encouraged—employees to speak their minds, do what they enjoyed doing, and generally behave at work as they behaved at home. This was driven, from the company's earliest days, by Sergey Brin and Larry Page, both educated in their formative years in Montessori schools that fostered free thought.

The tension over Project Maven was potentially even higher. Many of the scientists overseeing Google's deep learning research were fundamentally opposed to autonomous weapons, including both Geoff Hinton and the founders of DeepMind. That said, many of the executives at the highest rungs of the Google hierarchy very much wanted to work with the Defense Department. Eric Schmidt, the chairman of the Google board, was also chairman of the Defense Innovation Board, a civilian organization created by the Obama administration that aimed to accelerate the movement of new technologies from Silicon Valley into the Pentagon. At a recent meeting of the board, Schmidt had said there was "clearly a large gap" between Silicon Valley and the Pentagon and that the board's primary mission was to close this gap. The Google brass also saw military work as another way of bootstrapping the company's cloud business. Well behind the scenes, the company was already working with the Defense Department. The previous May, about a month after the launch of Project Maven, a team of Googlers met with officials at the Pentagon, and the next day the company applied for the government certifications needed to house military data on its own computer servers. But as Mattis discussed these technologies inside Google HQ three months later, he knew that navigating the relationship would require some delicacy.

Mattis said he'd already seen the power of the company's technology on the battlefield. After all, American adversaries were using Google Earth, an interactive digital map of the globe stitched together from satellite images, to identify mortar targets. But he was keen that the U.S. should up its game. Now, with Project Maven, the Defense Department wanted to develop AI that could not only read satellite images but also analyze video captured by drones much closer to the battlefield. Mattis complimented Google on its "industry-leading technology" as well as its "reputation for corporate responsibility." This, he said, was part of why he was there. He worried about the ethics of artificial intelligence. He said that the company should make the DoD "feel uncomfortable"—push back against its traditional attitudes. "Your ideas are welcome in the Department," he said.

Across the table, Pichai said Google often thought about the ethics of AI. Increasingly, bad actors were using this kind of technology, he said, so it was important for the good actors to stay ahead. Then Mattis asked if Google could encode some sort of moral or ethical rules into these systems—something the Googlers knew was far from a realistic option. Giannandrea, who oversaw Google's AI work, emphasized that these systems were ultimately dependent on the quality of their training data. But Kent Walker, the Google general counsel, put it differently. These technologies, he said, held tremendous potential to save lives.

By the end of September, little more than a month after Mattis visited Google headquarters, the company signed a contract for three years of work on Project Maven. It was worth between $25 million and $30 million, with $15 million set to be delivered in the first eighteen months. For Google, that was a small sum, and some of it would have to be shared with others involved in the contract, but the company was angling for something bigger. That same month, the Pentagon invited American companies to bid for what it called JEDI, short for Joint Enterprise Defense Infrastructure, a ten-year, $10 billion contract to supply the Defense Department with the cloud computing services needed to run its core technologies. The question was whether Google would publicize its involvement in Project Maven as it pushed for JEDI and, in the future, other government contracts.

Three weeks after Mattis visited Google headquarters, the Future of Life Institute released an open letter calling on the United Nations to ban what it called "killer robots," another way of describing autonomous weapons. "As companies building the technologies in Artificial Intelligence and Robotics that may be repurposed to develop autonomous weapons, we feel especially responsible in raising this alarm," the letter read. "Lethal autonomous weapons threaten to become the third revolution in warfare. Once developed, they will permit armed conflict to be fought at a scale greater than ever, and at timescales faster than humans can comprehend." It was signed by over a hundred people in the field. They included Elon Musk, who had so frequently warned against the threat of superintelligence, as well as Geoff Hinton, Demis Hassabis, and Mustafa Suleyman. For Suleyman, these were technologies that required a new kind of oversight. "Who's making the decisions that will one day affect billions of people on our planet? And who's involved in that judgment process?" he asks. "We need to significantly diversify the range of contributors to that decision making process, and that means involving regulators much earlier in the process—policy makers, civil society activists, and the people who we seek to serve with our technologies—getting them deeply involved in the creation of our products and the understanding of our algorithms."

That September, as Google prepared to sign its contract for Project Maven, the sales staff overseeing the agreement traded emails asking whether the company should publicize it. "Do we announce? Can we talk about the reward? What instructions do we give the govt?" one Googler wrote. "If we stay silent, we can't control the message. That won't go well for our brand." He ultimately argued that Google should go public with the news, and others agreed. "It will eventually get out," another Googler said. "Wouldn't it be best to release it on our own terms?" The discussion continued for several days, and along the way, someone roped in Fei-Fei Li.

Li applauded the contract. "It's so exciting that we're close to getting MAVEN! That would be a great win," she wrote. "What an amazing effort you've been leading! Thank you!" But she also urged extreme caution when promoting it. "I think we should do a good PR on the story of DoD collaborating with GCP from a vanilla cloud technology angle," she wrote, referring to the Google Cloud Platform. "But avoid at ALL COSTS any mention or implication of AI." She knew the press would question the ethics of the project, if only because Elon Musk had primed the pump:

> *Weaponized AI is probably one of the most sensitized topics in AI—if not THE most. This is red meat for the media to find all ways to damage Google. You probably heard Elon Musk and his comment about AI causing WW3. There is also a lot of media attention on AI weapons, international competition, and the potential of geopolitical tension for AI. Google is already battling with privacy issues when it comes to AI and data. I don't know what would happen if the media picked up the theme that Google was building AI weapons or AI technologies to enable weapons for the Defense Industry. Google Cloud has been building our theme for democratizing AI in 2017. And Diane and I have been talking about Humanistic AI for the enterprise. I would be super careful to protect these very positive images.*

Google did not announce the project, and it asked that the DoD refrain from announcing the project, too. Even company employees would have to learn about Project Maven on their own.

LOOMING over Highway 101—the eight-lane thoroughfare that cuts through the heart of Silicon Valley—Hangar One is one of the largest freestanding buildings on Earth. Built in the 1930s to house flying dirigibles for the U.S. Navy, this gargantuan steel barn is nearly two hundred feet tall, and it covers more than eight acres of land, room enough for six football fields. It is part of Moffett Field, a one-hundred-year-old military air base sitting between the cities of Mountain View and Sunnyvale. Moffett is owned by NASA, which runs a research center in the shadow of Hangar One, but the space agency leases most of the air base to Google. The company uses the old steel hangars to test balloons that could one day deliver Internet access from the skies, and for years its executives, including Larry Page and Sergey Brin and Eric Schmidt, flew their private planes in and out of Silicon Valley on its private runways.

The new Google Cloud headquarters sat at the southern edge of Moffett Field, a trio of buildings encircling a grass courtyard littered with lawn tables and chairs where Googlers ate their lunches each afternoon. One of the buildings housed what Google called the Advanced Solutions Lab, where the company explored custom technologies for its largest customers. On October 17 and 18, inside this building, company officials met with Undersecretary of Defense Patrick Shanahan and his staff to scope out Google's role in Project Maven. Like many others at the highest rungs of the Department of Defense, Shanahan saw the project as a first step toward something larger. At one point, he said: "Nothing in the DoD should ever be fielded going forward without built-in AI capability." At least to the Googlers who arranged the contract, it seemed like the company would be a vital part of this long haul.

But first, Google had to build software for what was called an "air gap" system—a computer (or network of computers) that was literally surrounded by an air gap, so it isn't connected to any other network. The only way to get data into such a system is through some sort of physical device, like a thumb drive. Apparently, the Pentagon would load its drone footage onto this system, and Google needed a way of accessing this data and feeding it into a neural network. The arrangement meant that Google wouldn't have control over the system or even have a good picture of how the system was used. In November, a team of nine Google engineers was assigned to build the software for this system, but they never did. Soon realizing what it was for, they refused to be involved in any way.

After the New Year, as word of the project began to spread across the company, others voiced concerns that Google was helping the Pentagon make drone strikes. In February, the nine engineers told their story in a post sent across the company on its internal social network, Google+. Like-minded employees supported the stance and hailed these engineers as the Gang of Nine. On the last day of the month, Meredith Whittaker, the product manager in the cloud group who founded the AI Now Institute at NYU, one of the highest-profile organizations dedicated to AI ethics, penned a petition. It demanded that Pichai cancel the Project Maven contract. "Google," it said, "should not be in the business of war."

During a Google town hall meeting the next day, executives told employees that the Maven contract topped out at only $9 million and that Google was building technology only for "nonoffensive" purposes. But the unrest continued to grow. The night of the town hall, another five hundred people signed Whittaker's petition. One thousand more signed the next day. In early April, after more than thirty-one hundred employees had signed, the *New York Times* published a story revealing what was going on, and, days later, the head of the cloud group invited Whittaker to join a panel discussion about the Maven contract during a company town hall meeting. She and two other Googlers, both pro-Maven, debated the issue three separate times, so that it could be broadcast live to three separate time zones around the globe.

In London, inside DeepMind, more than half of the employees signed Whittaker's petition, with Mustafa Suleyman playing a particularly prominent role in the protest. Google's Project Maven contract was an attack on his fundamental beliefs. He saw the protests inside Google as evidence that a European sensibility was spreading to the U.S., shifting the direction of even the biggest tech companies. A groundswell in Europe had spawned the General Data Protection Regulation, or GDPR, a law that forced these companies to respect data privacy. Now a groundswell inside Google was forcing the company to rethink its approach to military work. As the controversy roiled, Suleyman urged Pichai and Walker to finalize ethical guidelines that would formally define what Google would and would not build.

IN mid-May, a group of independent academics addressed an open letter to Larry Page, Sundar Pichai, Fei-Fei Li, and the head of the Google cloud business. "As scholars, academics, and researchers who study, teach about, and develop information technology, we write in solidarity with the 3100+ Google employees, joined by other technology workers, who oppose Google's participation in Project Maven," the letter read. "We wholeheartedly support their demand that Google terminate its contract with the DoD, and that Google and its parent company Alphabet commit not to develop military technologies and not to use the personal data that they collect for military purposes." It was signed by more than a thousand academics, including Yoshua Bengio and several of Li's colleagues at Stanford.

Li was caught between what her bosses wanted in industry and what her peers wanted in academia. Her predicament was indicative of the larger push and pull between the two worlds that collided in these years. A technology that academics tinkered with for decades was now a fundamental cog in some of the world's largest and most powerful companies. Its future was driven by the hunt for money as much as anything else. So deeply did many now feel about the issue that they even turned on Hinton for not being sufficiently vociferous about his concerns. "I lost a lot of respect for him," said Jack Poulson, a former Stanford professor who worked in Google's Toronto office, a few floors below Hinton. "He never said anything." But behind the scenes, Hinton personally urged Sergey Brin to cancel the contract.

After the open letter was published and Fei-Fei Li apparently received death threats on Chinese message boards, she told many people she feared for her safety, and she insisted that joining Project Maven was not her doing. "I was not involved in the decision to apply for the Maven contract or to accept it," she later said, before nodding to her email about Elon Musk and World War III. "My warning to the sales team was accurate." On May 30, the *New York Times* published a front-page story about the controversy that led with her email, and the protest inside the company grew louder. A few days later, Google executives told employees it would not renew the contract.

Google's ultimate decision was part of a larger pushback against government contracts. Employees at Clarifai also took issue with their work on Project Maven. One engineer quit the project immediately after the visit from the three military officers, and others left the company in the weeks and months that followed. At Microsoft and Amazon, employees protested against military and surveillance contracts. But these protests weren't nearly as effective. And even at Google, the groundswell eventually vanished. The company parted ways with most of those who had stood up against Maven, including Meredith Whittaker and Jack Poulson. Fei-Fei Li returned to Stanford. Though Google had dropped the contract, it was still pushing in the same direction. A year later, Kent Walker took the stage at an event in Washington alongside General Shanahan and said that the Maven contract was not indicative of the company's broader aims. "That was a decision focused on a discrete contract," he said, "not a broader statement about our willingness or history of working with the Department of Defense."

17

IMPOTENCE

"THERE ARE PEOPLE IN RUSSIA WHOSE JOB IT IS TO TRY TO EXPLOIT OUR SYSTEMS. SO THIS IS AN ARMS RACE, RIGHT?"

Mark Zuckerberg wore the same outfit every day: a pigeon-gray cotton T-shirt with a pair of blue jeans. He felt this gave him more energy to run Facebook, which he liked to call "a community," not a company or a social network. "I really want to clear my life to make it so that I have to make as few decisions as possible about anything except how to best serve this community," he once said. "There's actually a bunch of psychology theory that even making small decisions around what you wear, or what you eat for breakfast, or things like that, they kind of make you tired and consume your energy." But when he testified before Congress in April 2018, he wore a navy suit and a Facebook-blue tie. Some called it his "I'm sorry" suit. Others said his haircut, strangely high on his forehead, made him look like a penitent monk.

A month earlier, newspapers in the United States and Britain reported that the British start-up Cambridge Analytica had harvested private information from the Facebook profiles of more than 50 million people before using this data to target voters on behalf of the Trump campaign in the run-up to the 2016 election. The revelation let loose an avalanche of rebuke from the media, public advocates, and lawmakers that piled atop the criticism already aimed at Zuckerberg and Facebook over the past several months. Summoned to Capitol Hill, Zuckerberg endured ten hours of testimony over two days. He answered more than six hundred questions from nearly a hundred lawmakers on issues new and old, including the Cambridge Analytica data breach, Russian interference in the election, fake news, and the hate speech that frequently spread across Facebook, inciting violence in places like Myanmar and Sri Lanka. Zuckerberg apologized time and again, though he didn't always seem apologetic. In private as well as in public, Zuckerberg has an almost robotic demeanor, blinking his eyes unusually often and, from time to time, making an unconscious clicking sound at the back of his throat that seems liked some sort of glitch in the machine.

Halfway into Zuckerberg's testimony in the Senate on the first day, Republican John Thune, the senior senator from South Dakota, questioned the effect of Zuckerberg's apologies, saying the Facebook founder had spent fourteen years publicly apologizing for one egregious mistake after another. Zuckerberg acknowledged this, but he said Facebook was now realizing it should operate in a new way, that it needed to not just offer software for sharing information online but also work to actively police what was shared. "What I think we've learned now across a number of issues—not just data privacy, but also fake news and foreign interference in elections—is that we need to take a more proactive role and a broader view of our responsibility," he said. "It's not enough to just build tools. We need to make sure that they're used for good." Thune said he was glad Zuckerberg was getting the message. But he wanted to know exactly how Facebook would tackle what was a ridiculously difficult problem. His example was hate speech, which seemed simple but was not. Linguistically, it was often hard to define and sometimes quite subtle, exhibiting different characteristics in different countries.

In response, Zuckerberg rewound to the early days of Facebook, laying out some of the boilerplate he and his staff had prepared in the days before his testimony. When he launched Facebook out of his dorm room in 2004, he said, people could share whatever they wanted on the social network. Then, if someone flagged what was shared as inappropriate, the company would take a look and decide if it should be taken down. Over the decades, he acknowledged, this had become its own sprawling shadow operation, with more than twenty thousand contract employees working to review flagged content on a social network that juggled information from over 2 billion people. But, he said, artificial intelligence was changing what was possible.

Whereas researchers like Ian Goodfellow saw deep learning as something that could exacerbate the fake-news problem, Zuckerberg painted it as the solution. AI systems, he told Senator Thune, were already identifying terrorist propaganda with near perfect accuracy. "Today, as we sit here, 99 percent of the ISIS and Al Qaeda content that we take down on Facebook, our AI systems flag before any human sees it," he said. He acknowledged that other types of toxic content were harder to identify, including hate speech. But he was confident that artificial intelligence could solve the problem. In five to ten years, he said, it could even recognize the nuances of hate speech. What he didn't say was that even humans couldn't agree on what was and what was not hate speech.

TWO years earlier, in the summer of 2016, after AlphaGo defeated Lee Sedol and before Donald Trump defeated Hillary Clinton, Zuckerberg sat down at a conference table inside Building 20, the new centerpiece of the company campus in Menlo Park. Designed by Frank Gehry, it was a long, flat building on steel stilts that spanned more than four hundred thirty thousand square feet, seven times the size of a football field. The roof was its own Central Park, nine acres of grass and trees and gravel walkways where Facebookers sat or strolled whenever they liked. On the inside, it was one big open space holding over two thousand eight hundred employees, and filled with desks and chairs and laptops. If you stood in the right spot, you could see from one end all the way to the other.

Zuckerberg was holding a midyear company review. The leaders of each division would walk into the room where he was sitting, discuss their progress over the first six months of the year, and then walk out. That afternoon, the group overseeing the company's AI research walked in alongside Mike Schroepfer, the chief technology officer, and Yann LeCun gave the presentation, detailing their work with image recognition, translation, and natural language understanding. Zuckerberg didn't say much as he listened. Neither did Schroepfer. Then, when the presentation ended, the group walked out of the room, and Schroepfer lit into LeCun, telling him that nothing he'd said meant anything. "We just need something that shows we're doing better than other companies," he told LeCun. "I don't care how you do it. We just need to win a contest. Just start a contest we know we can win."

"Video. We can win video," said one of the colleagues standing over his shoulder.

"See?" Schroepfer barked at LeCun. "You can learn something."

Zuckerberg wanted the world to see Facebook as an innovator—as a rival to Google. This would help the company attract talent. And as the specter of antitrust rose over Silicon Valley, it would provide an argument against breaking the company into pieces—or so many at the company believed. The idea was that Facebook could show regulators that it was not just a social network, that it was more than just a connection among people, that it was a company building new technologies vital to the future of humanity. The Facebook AI lab was as much a PR opportunity as anything else. This was why Schroepfer had told a roomful of reporters that Facebook was building AI that could solve the game of Go, and that the big ideas behind the project would spread across the company. It was also why Zuckerberg and LeCun tried to preempt DeepMind's Go milestone several weeks later. But LeCun, the head of the Facebook lab, was not someone who chased the moonshot moments. He was not Demis Hassabis or Elon Musk. Having worked in the field for decades, he saw AI research as a much longer and much slower endeavor.

As it turned out, the big ideas Schroepfer described to that roomful of reporters were vitally important to the future of the company. But the future wouldn't prove as bright as he imagined it would be, the ideas weren't as big as they seemed, and they spread through the company in ways he never could have predicted.

After hiring dozens of top researchers for the Facebook AI lab, which spanned offices in New York and Silicon Valley, Schroepfer built a second organization charged with putting the lab's technologies into practice. This was called the Applied Machine Learning Team. In the beginning, it brought things like face recognition and language translation and automatic image captioning to the world's largest social network. But then its mission began to change. In late 2015, when Islamic militants killed one hundred and thirty people and wounded more than four hundred during coordinated attacks in and around Paris, Mark Zuckerberg sent an email asking the team what it could do to combat terrorism on Facebook. Over the next several months, they analyzed thousands of Facebook posts involving terrorist organizations that violated its policies, and came up with a system that could automatically flag new terrorist propaganda on its own. Then human contractors would review what it had highlighted and ultimately decide if it should be removed. It was this technology that Zuckerberg was pointing to when he told the Senate that Facebook's artificial intelligence could automatically identify content from ISIS and Al Qaeda. Others, however, questioned how sophisticated and nuanced such technology could be.

In November 2016, when Zuckerberg was still in denial about the role of Facebook in spreading fake news, Dean Pomerleau threw down the gauntlet. Thirty years after he built a self-driving car at Carnegie Mellon with help from a neural network, Pomerleau tweeted out what he called the "Fake News Challenge," betting $1,000 that no researcher could build an automated system that could separate the real from the fake. "I will give anyone 20:1 odds (up to $200 per entry; $1000 total) that they can't develop an automated algorithm that can distinguish real vs. fake claims on the internet," he wrote. He knew that current technology was not up to the task, which required a very subtle breed of human judgment. Any AI technology that could reliably identify fake news would have passed a much larger milestone. "It would mean AI has reached human-level intelligence," he said. He also knew that fake news was in the eye of the beholder. Separating the real from the fake was a matter of opinion. If humans couldn't agree on what was and what was not fake news, how could they train machines to recognize it? News was, inherently, a tension between objective observation and subjective judgment. "In many cases," Pomerleau said, "there is no right answer." Initially, there was a flurry of activity in response to his challenge. Nothing came of it.

A day after Pomerleau issued his challenge, as Facebook continued to deny there was a problem, the company held a press roundtable at its corporate headquarters in Menlo Park. Yann LeCun was there, and a reporter asked him if AI could detect fake news and other toxic content that spread so rapidly across the social network, including violence in live video. Two months earlier, a man in Bangkok had hung himself while broadcasting a live Facebook video. LeCun responded with an ethical conundrum. "What's the trade-off between filtering and censorship? Freedom of experience and decency?" he said. "The technology either exists or can be developed. But then the question is how does it make sense to deploy it? And this isn't my department."

As pressure mounted from outside the company, Schroepfer started shifting resources inside his Applied Machine Learning Team in an effort to clean up toxic activity across the social network, from pornography to fake accounts. By the middle of 2017, the detection of unwanted content accounted for more of the team's work than any other task. Schroepfer called it the "clear No. 1 priority." At the same time, the company continued to expand the number of human contractors who reviewed content. AI was not enough.

So when Zuckerberg testified before Congress after the Cambridge Analytica data leak he had to acknowledge that Facebook's monitoring systems still required human help. They could flag certain kinds of images and text, like a nude photo or a piece of terrorist propaganda, but once they had done that, human curators—huge farms of contractors working mostly overseas—had to step in to review each post and decide whether it should be taken down.

However accurate the AI tools might be in very specific situations, they still lacked the flexibility of human judgment. They struggled, for example, to distinguish a pornographic image from a photo of a mother breastfeeding a baby. And they weren't dealing with a static situation: As Facebook built systems that could identify an increasingly diverse array of toxic content, new kinds of material would appear on the social network that these systems were not trained to identify. When Senator Dianne Feinstein, the Democrat from California, asked Zuckerberg how he was going to stop foreign actors from interfering with U.S. elections, he again pointed to AI. But he acknowledged that the situation was complicated. "We've deployed new AI tools that do a better job of identifying fake accounts that may be trying to interfere in elections or spread misinformation. The nature of these attacks, though, is that, you know, there are people in Russia whose job it is to try to exploit our systems," he said. "So this is an arms race, right?"

IN March 2019, a year after Zuckerberg testified on Capitol Hill, a gunman killed fifty-one people at two mosques in Christchurch, New Zealand, while live-streaming the attack on Facebook. It took the company an hour to remove the video from its social network, and in that hour, it spread across the Internet. Days later, Mike Schroepfer sat down with two reporters in a room at Facebook headquarters to discuss the company's efforts to identify and remove unwanted content with help from artificial intelligence. For half an hour, while drawing diagrams on a whiteboard in colored marker, he showed how the company was automatically identifying ads for marijuana and ecstasy. Then the reporters asked about the shootings in Christchurch. He paused for nearly sixty seconds before tearing up. "We're working on this right now," he said. "It won't be fixed tomorrow. But I do not want to have this conversation again six months from now. We can do a much, much better job of catching this."

Over the course of several interviews, he teared up time and again as he discussed the scale and the difficulty of Facebook's task and the responsibilities that came with it. Through it all, he insisted that AI was the answer, that as time went on it would reduce the company's struggles and eventually turn a Sisyphean task into a manageable situation. But when pressed he acknowledged that the problem would never vanish entirely. "I do think there's an endgame here," he said. But "I don't think it's 'everything's solved,' and we all pack up and go home."

In the meantime, building this artificial intelligence was a very human endeavor. When the Christchurch video appeared on Facebook, the company's systems didn't flag it because it didn't look like anything these systems had been trained to recognize. It looked like a computer game, with a first-person point of view. Facebook had trained systems to identify graphic violence on its site, using images of dogs attacking people, people kicking cats, people hitting other people with baseball bats. But the New Zealand video was different. "None of those look a lot like this video," Schroepfer said. Like others on his team, he'd watched the video several times in an effort to understand how they could build a system that could identify it automatically. "I wish I could unsee it," he said.

As marijuana ads appeared on the social network, Schroepfer and his team built systems that could identify them. Then new and different activity appeared, and they built new ways of identifying this new content, as the cycle continued. Amid it all, researchers were building systems that could generate misinformation on their own. This included GANs and related techniques for image generation. It also included a technology developed at DeepMind called WaveNet that could generate realistic sounds, even help duplicate someone's voice, like Donald Trump's or Nancy Pelosi's.

This was evolving into a game of AI versus AI. As another election approached, Schroepfer launched a contest that urged researchers from across industry and academia to build AI systems that could identify deepfakes, fake images generated by other AI systems. The question was: Which side would win? For researchers like Ian Goodfellow, the answer was obvious. The misinformation would win. GANs, after all, were designed as a way of building a creator that could fool any detector. It would win even before the game was played.

A few weeks after Facebook unveiled its contest, another reporter asked Yann LeCun, once again, if artificial intelligence could stop fake news. "I am not sure anyone has the technology to access the truthfulness of a piece of news," he said. "Truthfulness, particularly when it comes to political questions, is very much in the eye of the beholder." Even if you could build a machine that could do a reasonable job of this, he added, many people would say that the technologists who built it were biased and they would complain the data used to train it was biased and they just wouldn't accept it. "Even if the technology existed," he said, "it may not be a good idea to actually deploy it."

PART FOUR

HUMANS ARE UNDERRATED

18

DEBATE

"HOWEVER LONG THE RAPID PROGRESS CONTINUES, GARY WILL STILL ARGUE IT IS ABOUT TO END."

The centerpiece of the Google year is a conference called I/O. It's named for an old computer acronym that stands for Input/Output. Each May, thousands make the pilgrimage to Mountain View for this corporate extravaganza, traveling from across Silicon Valley and more distant parts of the tech industry just so they can spend the better part of three days learning about the company's newest products and services. Google holds its annual conference keynote at the Shoreline Amphitheatre, a twenty-two-thousand-seat concert venue whose circus-tent-like spires rise above the grassy hills across the street from company headquarters. Over the decades, everyone from the Grateful Dead to U2 to the Backstreet Boys played the Shoreline. Now it's where Sundar Pichai takes the stage to tell thousands of software developers about the myriad technologies emerging from his increasingly diversified company. On the opening day of the conference in the spring of 2018, wearing a forest-green fleece zipped up over a bright white T-shirt, Pichai told the crowd that the company's talking digital assistant could make its own phone calls.

Thanks to the methods pioneered by Geoff Hinton and his students in Toronto, the Google Assistant could recognize spoken words nearly as well as people could. Thanks to WaveNet, the speech-generation technology developed at DeepMind, it *sounded* more human, too. Then, standing onstage at the Shoreline, Pichai unveiled a new refinement. The Google Assistant, he told his audience, could now call a restaurant and make a reservation. It did this in the background, through Google's computer network. You could ask the Assistant to book a table for dinner, and then, while you did something entirely different, like take out the garbage or water the lawn, the Assistant would trigger an automated phone call to your restaurant of choice from somewhere inside a Google data center. Pichai played an audio recording of one of these calls, a chat between the Assistant and a woman who answered the phone at an unnamed restaurant.

"Hi, may I help you?" the woman said, with a heavy Chinese accent.

"Hi, I would like to reserve a table for Wednesday, the seventh," Google Assistant said.

"For seven people?" the woman asked, as giggles rippled across the amphitheater.

"Um, it is for four people," Google Assistant said.

"Four people. When? Today? Tonight?" the woman at the restaurant said, as the giggles grew louder.

"Um, next Wednesday at six p.m."

"Actually we reserve for upwards of five people. For four people, you can come."

"How long is the wait usually to, uh, be seated?"

"For when? Tomorrow? Or weekend?"

"For next Wednesday, uh, the seventh."

"Oh, no, it is not too busy. You can come for four people, OK?"

"Oh, I gotcha. Thanks."

"Yep. Bye-bye," the woman said, as cheers and whoops erupted from Pichai's audience.

As Pichai explained, this new technology was called Duplex, and it was the result of several years of progress in a wide range of AI technologies, including speech recognition and speech generation as well as natural language understanding—the ability to not just recognize and generate spoken words but to truly understand the way language is used. For those in the audience, Pichai's demo was shockingly powerful. Then he played them a second in which the system made a haircut appointment at a local salon. The applause came when the woman at the spa said "Give me one second" and Duplex responded with an "Mmm-hmmm." Duplex could respond not just with the right words but with the right noises—the right verbal cues. In the days that followed, many pundits complained that Google Duplex was so powerful, it was unethical. It was actively deceiving people into thinking it was human. Google agreed to tweak the system so that it always disclosed it was a bot, and soon it released the tool in various parts of the United States.

But for Gary Marcus, the technology was not what it seemed.

Days after Pichai gave his demo at the Shoreline Amphitheatre, Marcus, a professor of psychology at NYU, published an editorial in the *New York Times* that aimed to put Google Duplex in its place. "Assuming the demonstration is legitimate, that's an impressive (if somewhat creepy) accomplishment. But Google Duplex is not the advance toward meaningful AI that many people seem to think." The trick, he said, was that this system was operating in a tiny domain: restaurant reservations and hair salon appointments. By keeping the scope small—limiting the possible responses on each side of the conversation—Google could fool people into believing a machine was a human. That was very different from a system that could step outside those bounds. "Schedule hair salon appointments? The dream of artificial intelligence was supposed to be grander than this—to help revolutionize medicine, say, or to produce trustworthy robot helpers for the home," he wrote. "The reason Google Duplex is so narrow in scope isn't that it represents a small but important first step toward such goals. The reason is that the field of AI doesn't yet have a clue how to do any better."

GARY Marcus came from a long line of thinkers who believe in the importance of nature, not just nurture. They're called nativists, and they argue that a significant portion of all human knowledge is wired into the brain, not learned from experience. This is an argument that has spanned centuries of philosophy and psychology, running from Plato to Immanuel Kant to Noam Chomsky to Steven Pinker. The nativists stand in opposition to the empiricists, who believe that human knowledge comes mostly from learning. Gary Marcus studied under Pinker, the psychologist, linguist, and popular science author, before building his own career around the same fundamental attitude. Now he was wielding his nativism in the world of artificial intelligence. He was the world's leading critic of neural networks, a Marvin Minsky for the age of deep learning.

Just as he believed that knowledge was wired into the human brain, he believed that researchers and engineers had no choice but to wire knowledge into AI. He was sure machines couldn't learn everything. Back in the early 1990s, he and Pinker had published a paper showing that neural networks couldn't even learn the language skills that very young children learn, like how to recognize everyday verbs in the past tense. Twenty years later, in the wake of AlexNet, when the *New York Times* published a front-page story on the rise of deep learning, he responded with a column for the *New Yorker* arguing that the change was not as big as it seemed. The techniques espoused by Geoff Hinton, he said, were not powerful enough to understand the basics of natural language, much less duplicate human thought. "To paraphrase an old parable, Hinton has built a better ladder," he wrote. "But a better ladder doesn't necessarily get you to the moon."

The irony was that not long after, Marcus cashed in on the deep learning craze. In the first days of 2014, hearing that DeepMind had sold itself to Google for $650 million, and thinking, "I can do that," he phoned an old friend named Zoubin Ghahramani. They'd met more than twenty years earlier, when they were both graduate students at MIT. Marcus was there to study cognitive science, and Ghahramani was part of a program that bridged the gap between computer science and neuroscience. They became friends because they shared a birthday, celebrating their twenty-first at Marcus's apartment on Magazine Street in Cambridge. After finishing his PhD, Ghahramani took a path not unlike many of the AI researchers who were now working for Google and Facebook and DeepMind. He did a postdoc under Geoff Hinton in Toronto, then followed him to the Gatsby Unit at University College London. But Ghahramani was among those who eventually moved away from neural network research, embracing ideas that he saw as more elegant, more powerful, and more useful. So, after DeepMind sold itself to Google, Marcus convinced Ghahramani they should create their own start-up around the idea that the world needed more than just deep learning. They called it Geometric Intelligence.

They hired about a dozen artificial intelligence researchers from universities across the U.S. Some specialized in deep learning.

Others, including Ghahramani, specialized in other technologies. Marcus was not unaware of the powers of the technology, and he certainly understood the hype that surrounded it. After founding their start-up in the summer of 2015, he and Ghahramani installed their team of academics at a small office in downtown Manhattan where NYU incubated start-ups. Marcus joined them, while Ghahramani stayed in Britain. Little more than a year later, after conversations with many of the largest tech companies, from Apple to Amazon, they sold their start-up to Uber, the rapidly growing ride-hailing company that aspired to build self-driving cars. The start-up's dozen researchers promptly moved to San Francisco, becoming the Uber AI Labs. Marcus moved with the lab, while Ghahramani stayed in the UK. Then, four months later, without much explanation, Marcus left the company and returned to New York, resuming his role as the world's leading critic of deep learning. He was not an AI researcher. He was someone dedicated to his own set of intellectual ideas. One colleague called him a "lovable narcissist." Back in New York, he started writing a book arguing, once again, that machines could learn only so much on their own, and he began building a second company based on the same premise. He also challenged the likes of Hinton to a public debate over the future of artificial intelligence. Hinton did not accept.

But in the fall of 2017, Marcus debated LeCun at NYU. Organized by the NYU Center for Mind, Brain, and Consciousness—a program that combined psychology, linguistics, neuroscience, computer science, and more—the debate matched nature against nurture, nativism against empiricism, "innate machinery" against "machine learning." Marcus was the first to speak, arguing that deep learning was not capable of much more than simple tasks of perception, like recognizing objects in images or identifying spoken words. "If neural networks have taught us anything, it is that pure empiricism has its limits," he said. On the long road to artificial intelligence, he explained, deep learning had taken only a few tiny steps. Beyond tasks of perception (like image and speech recognition) and media generation (like GANs), its biggest accomplishment was solving Go, and Go was merely a game, a contained universe, with a carefully defined set of rules. The real world was almost infinitely more complex. A system trained to play Go, Marcus liked to say, was useless in any other situation. It was not intelligent because it could not adapt to entirely new situations. And it certainly couldn't handle one of the key products of human intelligence: language. "Sheer bottom-up statistics hasn't gotten us very far on an important set of problems—language, reasoning, planning, and common sense—even after sixty years of neural network research, even after we have vastly better computation, much more memory, much better data," he told his audience.

The trouble, he explained, is that neural networks do not learn in the way the human brain learns. Even when mastering tasks that neural networks cannot master, the brain doesn't require the enormous amounts of data that deep learning does. Children, including newborn babies, can learn from tiny amounts of information, sometimes just one or two good examples. Even children growing up in homes where their parents showed no interest in their development and education learn the nuance of spoken language just by listening to what is happening around them. A neural network, he argued, needs not only thousands of examples. It needs someone to carefully label each one. What this showed was that artificial intelligence wouldn't happen without more of the stuff that nativists called "innate machinery": the enormous amount of knowledge that they believed was baked into the human brain. "Learning is only possible because our ancestors have evolved machinery for representing things like space, time, and enduring objects," Marcus said. "My prediction—and it is only a prediction; I won't be able to prove it—is that AI will work much better when we learn to incorporate similar information into AI." In other words, he believed there were many things AI could never learn on its own. These things would have to be hand-coded by engineers.

As a confirmed nativist, Marcus had an ideological agenda. Working to build a new AI start-up around the idea of innate machinery, he also had an economic agenda. The debate with Yann LeCun at NYU was the beginning of a concerted campaign to show the worldwide community of AI researchers, the tech industry, and the general public that deep learning was far more limited than it seemed. In the early months of 2018, he published what he called a trilogy of papers critiquing deep learning and, in particular, the feats of AlphaGo. Then he pitched his critique in the popular press, with one story appearing on the cover of *Wired* magazine. All this would eventually lead to a book he titled *Rebooting AI*—and a new start-up that aimed to exploit what he saw as a hole in the global effort to build artificial intelligence.

LeCun was bemused by it all. As he told the audience at NYU, he agreed that deep learning alone could not achieve true intelligence, and he had never said it would. He agreed that AI would require innate machinery. After all, a neural network was innate machinery. Something had to do the learning. He was measured and even polite during the debate. But his tone changed online. When Marcus published his first paper questioning the future of deep learning, LeCun responded with a tweet: "The number of valuable recommendations ever made by Gary Marcus is exactly zero."

Marcus was not alone. Many were now pushing back against the endless wave of hype emerging from the industry and the press around the words "artificial intelligence." Facebook was at the forefront of the deep learning revolution and it held up the technology as an answer to its most pressing problems. It was increasingly obvious, however, that this was, at best, a partial solution. For years, companies like Google and Uber had promised that self-driving cars would soon be on the roads, shuttling everyday people across cities in the United States and abroad. But even the popular press began to realize these claims were grossly exaggerated. Although deep learning had significantly improved their ability to recognize people, objects, and signs on the road—and accelerated their ability to predict events and plan routes forward—self-driving cars were still a long way from dealing with the chaos of the daily commute with the same agility as people. Though Google had promised a ride-hailing service in Phoenix, Arizona, by the end of 2018, that did not happen. Deep learning for drug discovery, an area that seemed to hold such promise after George Dahl and his Toronto collaborators won the Merck competition, turned out to be a far more complicated proposition than it seemed. Not long after coming to Google, Dahl moved away from the idea. "The problem is that the part of the drug discovery pipeline that we're most able to help is not the most important part of the pipeline," he says. "It is not the part that makes it cost $2 billion to bring a molecule to market." Oren Etzioni, the former University of Washington researcher who oversaw the Allen Institute for Artificial Intelligence, often said that for all the hype around deep learning, AI could not even pass an eighth-grade science test.

When Yann LeCun had unveiled Facebook's new Paris lab in June 2015, he said: "The next big step for deep learning is natural language understanding, which aims to give machines the power to understand not just individual words but entire sentences and paragraphs." That was the goal of the wider community—the next big step beyond image and speech recognition. A machine that could understand the natural way people write and talk—and even carry on a conversation—had been the ultimate goal for AI research since the 1950s. But by the end of 2018 many felt such confidence to be misplaced.

Near the end of their debate, as Marcus and LeCun took questions from their audience, a woman in a yellow blouse stood up and asked LeCun why progress had stalled with natural language.

"Nothing as revolutionary as object recognition has happened," she said.

"I don't entirely agree with your premise," LeCun said. "There has—"

Then she cut him off, saying: "What is your example?"

"Translation," he said.

"Machine translation," she said, "is not necessarily language understanding."

AROUND the same time as this debate, researchers at the Allen Institute for Artificial Intelligence unveiled a new kind of English test for computer systems. It tested whether machines could finish sentences like this:

Onstage, a woman takes a seat at the piano. She

a. sits on a bench as her sister plays with the doll.

b. smiles with someone as the music plays.

c. is in the crowd, watching the dancers.

d. nervously sets her fingers on the keys.

Machines did not do very well. Whereas humans answered more

than 88 percent of the test questions correctly, a system built by the Allen Institute topped out at around 60 percent. Other machines did considerably worse. Then, about two months later, a team of Google researchers, led by a man named Jacob Devlin, unveiled a system they called BERT. When BERT took the test, it could answer just as many questions as a human could. And it wasn't actually designed to take the test.

BERT was what researchers call a "universal language model." Several other labs, including the Allen Institute and OpenAI, had been working on similar systems. Universal language models are giant neural networks that learn the vagaries of language by analyzing millions of sentences written by humans. The system built by OpenAI analyzed thousands of self-published books, including romance, science fiction, and mysteries. BERT analyzed the same vast library of books as well as every article on Wikipedia, spending days poring over all this text with help from hundreds of GPU chips.

Each system learned a very specific skill by analyzing all this text. OpenAI's system learned to guess the next word in a sentence. BERT learned to guess missing words *anywhere* in a sentence (such as "The man ____ the car because it was cheap"). But in mastering these specific tasks, each system also learned about the general way that language is pieced together, the fundamental relationships between thousands of English words. Then researchers could readily apply this knowledge to a wide range of other tasks. If they fed thousands of questions and their answers into BERT, it learned to answer other questions on its own. If they fed reams of running dialogue into OpenAI's system, it could learn to carry on a conversation. If they fed it thousands of negative headlines, it could learn to recognize a negative headline.

BERT showed that this big idea could work. It could handle the "common sense" test from the Allen Institute. It could also handle a reading comprehension test where it answered questions about encyclopedia articles. What is carbon? Who is Jimmy Hoffa? In another test, it could judge the sentiment of a movie review—whether it was positive or negative. It was hardly perfect in these situations, but it instantly changed the course of natural language research, accelerating progress across the field in a way that had never been possible before. Jeff Dean and Google open-sourced BERT and soon trained it in more than a hundred languages. Others built even larger models, training them on ever larger amounts of data. As a kind of inside joke among researchers, these systems were typically named after *Sesame Street* characters: ELMO, ERNIE, BERT (short for Bidirectional Encoder Representations from Transformers). But this belied their importance. Several months later, using BERT, Oren Etzioni and the Allen Institute built an AI system that could pass an eighth-grade science test—and a twelfth-grade test, too.

In the wake of BERT's release, the *New York Times* published a story on the rise of universal language models, explaining how these systems could improve a wide range of products and services, including everything from digital assistants like Alexa and the Google Assistant to software that automatically analyzed documents inside law firms, hospitals, banks, and other businesses. And it explained the concern that these language models could lead to more powerful versions of Google Duplex, bots designed to convince the world that they're human. The story also quoted Gary Marcus saying that the public should be skeptical that these technologies will continue to improve so rapidly because researchers tend to focus on the tasks they can make progress on and avoid the ones they can't. "These systems are still a really long way from truly understanding running prose," he said. When Geoff Hinton read this, he was amused. The quote from Gary Marcus would prove useful, he said, because it could slot into any story written about AI and natural language for years to come. "It has no technical content so it will never go out of date," Hinton said. "And however long the rapid progress continues, Gary will still argue it is about to end."

19

AUTOMATION

"IF THE ROOM LOOKED LIKE
THINGS HAD GONE CRAZY, WE WERE
ON THE RIGHT TRACK."

One afternoon in the fall of 2019, on the top floor of OpenAI's three-story building in San Francisco's Mission District, a hand was poised near the window, palm up, fingers outstretched. It looked a lot like a human hand, except it was made of metal and hard plastic and wired for electricity. Standing nearby, a woman scrambled a Rubik's Cube and placed it in the palm of this mechanical hand. The hand then began to move, gently turning the colored tiles with its thumb and four fingers. With each turn, the cube teetered at the end of its fingertips, nearly tumbling to the floor. But it didn't. As the seconds passed, the colors began lining up, red next to red, yellow next to yellow, blue next to blue. About four minutes later, the hand twisted the cube one last time, unscrambling the tiles. A small crowd of onlooking researchers let out a cheer.

Led by Wojciech Zaremba, the Polish researcher stolen from under Google and Facebook as OpenAI was founded, they'd spent more than two years working toward this eye-catching feat. In the past, many others had built robots that could solve a Rubik's Cube. Some devices could solve it in less than a second. But this was a new trick. This was a robotic hand that moved like a human hand, not specialized hardware built solely for solving Rubik's Cubes. Typically, engineers programmed behavior into robots with painstaking precision, spending months defining elaborate rules for each tiny movement. But it would take decades, maybe even centuries, for engineers to individually define each piece of behavior a five-fingered hand would need to solve a Rubik's Cube. Zaremba and his team had built a system that could learn this behavior on its own. They were part of a new community of researchers who believed robots could learn practically any skill inside virtual reality before applying it in the real world.

They began the project by creating a digital simulation of both the hand and the Cube. Inside this simulation, the hand learned through constant trial and error, spending the equivalent of ten thousand years spinning the tiles this way and that, discovering which tiny movements worked and which didn't. And over those ten thousand virtual years, the simulation was constantly changing. Zaremba and his team repeatedly changed the size of the fingers and the colors on the Cube and the amount of friction between the tiles and even the colors in the empty space behind the Cube. This meant that when they transferred all this virtual experience to the real hand in the real world, it could deal with the unexpected. It could handle the uncertainty that humans manage to accommodate so easily in the physical world but machines, so often, cannot. By the fall of 2019, OpenAI's robotic hand could solve a Rubik's

Cube with two fingers tied together or while wearing a rubber glove or with someone nudging the Cube out of place with the nose of a stuffed toy giraffe.

FROM 2015 to 2017, Amazon ran an annual contest for roboticists. In its final year, seventy-five academic labs entered this international competition, each working to build a robotic system that could solve the problem Amazon most needed to solve inside its global network of warehouses: *picking.* As bins filled with piles of retail products moved through Amazon's giant warehouses, human workers picked through the piles and sorted the products into the right cardboard boxes before they were shipped across the country. Amazon wanted robots to do the job. Ultimately, if this task could be automated, it would be less expensive. But robots weren't really up to it. So Amazon ran a contest, offering $80,000 to the academics who could come closest to cracking the problem.

In July of 2017, sixteen finalists from ten countries traveled to Nagoya, Japan, for the last round of the competition. Each team, which had spent a year preparing for the contest, was presented with a bin filled with thirty-two different items, sixteen of which they knew about ahead of time and sixteen of which they didn't—including bottles of Windex, ice trays, tennis ball cans, boxes of Magic Markers, and rolls of electrical tape. The challenge was to pick at least ten items in fifteen minutes. The winning robotic arm belonged to a team from a lab in Australia—the Australian Centre for Robotic Vision. But its performance was unimpressive by human standards. It picked the wrong item about ten percent of the time, and it could handle only about a hundred and twenty products an hour, little more than a quarter of what humans can do.

If the contest revealed anything, it was the difficulty of this task for even the most agile of robots. But it also showed what the industry needed: Amazon—and companies like Amazon—were desperate for picking robots that actually worked. The solution, as it turned out, was already brewing inside both Google and OpenAI.

After building a medical team inside Google Brain, Jeff Dean built a robotics team, too. One of his first hires was a young researcher from the University of California–Berkeley named Sergey Levine. Levine had grown up in Moscow, where both his parents worked as engineers on the Buran project, the Soviet version of the Space Shuttle. He moved to the United States while still in elementary school, and when he started his PhD, he was not an AI researcher. He specialized in computer graphics, exploring ways of building a more realistic breed of animation—virtual people that behaved more like real people. Then deep learning came of age, and his research accelerated. Through the same sort of techniques that DeepMind researchers used in building systems that learned to play old Atari games, Levine's animated figures could learn to move like people. Then came a new revelation. Watching these animated humanoids learn to move like he moved, Levine realized that physical humanoids could learn the same movements in much the same way. If he applied these machine learning techniques to robots, they could master entirely new skills on their own.

When he joined Google in 2015, Levine already knew Ilya Sutskever, another Russian émigré, and Sutskever introduced him to Alex Krizhevsky, who started working with the new robotics group. If he ever ran into a problem, Levine would ask Krizhevsky for help, and Krizhevsky's advice was always the same: Collect more data. "If you have the data and it's the right kind of data," Krizhevsky would say, "then just get more of it." So Levine and his team built what they called the Arm Farm.

In a big open room in a building down the street from the Brain lab, they set up a dozen robotic arms—six arms against one wall, six against the other. They were simpler than the arm that would later solve the Rubik's Cube inside OpenAI. The hands on these arms were not quite hands. They were "grippers" that could grab and lift objects between two viselike fingers. That fall, Levine and his team perched each arm over a bin of random stuff—toy blocks, blackboard erasers, tubes of lipstick—and trained them to pick up whatever was there. The arms learned by repeated trial and error, trying and failing to pick up what was in front of them until they discovered what worked and what didn't. It was a lot like the way DeepMind's systems learned to play Space Invaders and Breakout, except that it happened in the real world, with physical objects.

At first, this created chaos. "It was a mess," Levine says, "a really horrible mess." Per Krizhevsky's advice, they kept the robots running around the clock, and though they had set up a camera that let them peek into the room at night and on weekends, sometimes the chaos took over. They would walk into the lab on a Monday morning to find the floor strewn with stuff, like a children's playroom. One morning, they walked in and one bin was covered with what looked like splattered blood. A cap had come off a lipstick, and the arm spent the night trying and failing to pick it up. But this was exactly what Levine wanted to see. "It was wonderful," he says. "If the room looked like things had gone crazy, we were on the right track." As the weeks passed, these arms learned to pick up whatever was sitting in front of them with an almost gentle touch.

It marked the beginning of a widespread effort to apply deep learning to robotics, spanning labs inside many top universities as well as Google and OpenAI. The following year, using the same kind of reinforcement learning, Levine and his team trained other arms to open doors on their own (provided the doorknobs could be gripped with two fingers). In early 2019, the lab unveiled a robotic arm that learned to pick up random objects and then gently toss them into a tiny bin several feet away. The training took only fourteen hours, and the arm learned to toss items into the right bin about 85 percent of the time. When the researchers themselves tried the same task, they topped out at 80 percent. But as this work progressed, OpenAI took a different tack.

ELON Musk and the other founders of OpenAI saw their lab as an answer to DeepMind. From the beginning, their aim was to reach for enormously lofty goals that were easy to measure, easy to understand, and guaranteed to grab attention, even if they didn't actually do anything practical. After setting up their lab above a tiny chocolate factory in the San Francisco Mission District, researchers like Zaremba spent weeks walking around this old, rapidly gentrifying Hispanic neighborhood, debating what lofty goal they should chase. They eventually settled on two: a machine that could beat the world's best players at a three-dimensional video game called Dota, and a five-fingered robotic hand that could solve the Rubik's Cube. With their robotic hand, Wojciech Zaremba and his team used the same algorithmic technique as their counterparts at Google. But they moved the training into virtual reality, building a system that learned to solve the Rubik's Cube through centuries of trial and error in the digital world. Training systems in the physical world, they believed, would be far too expensive and time-consuming as the tasks grew more complex.

Like the lab's effort to master Dota, the Rubik's Cube project would require a massive technological leap. Both projects were also conspicuous stunts, a way for OpenAI to promote itself as it sought to attract the money and the talent needed to push its research forward. The techniques under development at labs like OpenAI were expensive—both in equipment and in personnel—which meant that eye-catching demonstrations were their lifeblood.

This was Musk's stock in trade: drawing attention to himself and what he was doing. For a while, it worked with OpenAI, too, as the lab hired some of the biggest names in the field. This included Sergey Levine's former advisor at University of California–Berkeley, a six-foot-two-inch, slick-bald Belgian roboticist named Pieter Abbeel. Abbeel's OpenAI signing bonus was $100,000, and his salary for just the last six months of 2016 alone was $330,000. Three of Abbeel's former students also joined OpenAI as it accelerated efforts to challenge Google Brain and Facebook and, particularly, DeepMind. Then reality caught up with both Musk and his new lab.

Ian Goodfellow, the GANfather, left and returned to Google. Musk himself poached a top researcher from the lab, lifting a computer vision expert named Andrej Karpathy out of OpenAI and installing him as the head of artificial intelligence at Tesla so he could lead the company's push into self-driving cars. Then Abbeel and two of his students left to create their own robotics start-up. And in February 2018, Musk left, too. He said he left to avoid conflicts of interest—meaning that his other businesses were now competing for the same talent as OpenAI—but he was also facing a crisis at Tesla, as crippling slowdowns inside its factories threatened to put the car company out of business. The irony was that, as Musk complained later that year, the robotic machines that helped manufacture the electric cars inside his Tesla factories weren't as nimble as they seemed. "Excessive automation at Tesla was a mistake," he said. "Humans are underrated."

As Sam Altman took the reins at OpenAI, the lab needed to attract new talent—and it needed money. Though investors had committed a billion dollars to the nonprofit when it was founded, only a fraction of that money actually arrived, and the lab now needed far more, not just to attract talent but also to pay for the massive amounts of computing power needed to train its systems. So Altman re-formed the lab as a for-profit company and went after new investors. The idealistic vision of a lab free from corporate pressures that he and Musk preached when they unveiled the lab in 2015 did not last even four years. All this was why the Rubik's Cube project was so important to the lab's future. It was a way for OpenAI to promote itself. The rub was that this kind of incredibly difficult but ultimately nonpractical project was not what Abbeel and others wanted to work on. He wasn't interested in generating hype. He wanted to build useful technology. This was why he and two of his former Berkeley students, Peter Chen and Rocky Duan, left the lab to found a start-up called Covariant. Their new company was committed to the same techniques OpenAI was exploring, except the aim was to apply them in the real world.

By 2019, as researchers and entrepreneurs recognized what Amazon and the rest of the world's retailers needed in their warehouses, the market was flooded with robotic-picking start-ups, some of them employing the sort of deep learning methods under development at Google Brain and OpenAI. Pieter Abbeel's company, Covariant, wasn't necessarily one of them. It was designing a system for a much wider range of tasks. But then, two years after the Amazon Robotics Challenge, an international robotics maker called ABB organized its own contest, this one behind closed doors. Covariant decided to join.

Nearly twenty companies entered this new contest, which involved picking about twenty-five different products, some of which the companies were told about ahead of time, some of which they weren't. Items in the mix included bags of gummy bears and clear bottles filled with soap or gel, which were particularly hard for robots to pick up because they reflected the light in so many unexpected ways. Most companies failed the test outright. A few handled most of the tasks but failed the more difficult scenarios, like having to pick up old audio CDs that reflected the light from above and sometimes sat on end, leaning against the side of the bin.

Abbeel and his colleagues had initially wondered whether to participate at all, given that they hadn't really built their system for picking. But their new system could learn. Over several days, they trained it on a wide array of new data, reaching a point, when ABB visited their lab in Berkeley, where their robotic arm handled every task as well as, or even better than, a human. Its one mistake was accidentally dropping a gummy-bear bag. "We were trying to find weaknesses," said Marc Segura, managing director of service robotics at ABB. "It is easy to reach a certain level on these tests, but it is super difficult not to show any weaknesses."

As the company developed this technology, further funding was required, so Abbeel decided he would ask the biggest names in AI. Yann LeCun visited the lab in Berkeley and agreed to invest after pouring a few dozen empty plastic bottles into a bin and watching the arm pick them without a hitch. Yoshua Bengio declined to invest. Though he had taken only part-time gigs with the big tech companies, he said that he had more money than he could ever spend and that he preferred to focus on his own research. But Geoff Hinton invested. He believed in Abbeel. "He is good," Hinton says. "And that's very surprising. After all, he's Belgian."

That fall, a German electronics retailer moved Abbeel's technology into a warehouse on the outskirts of Berlin, where it started picking and sorting switches, sockets, and other electrical parts from blue crates as they moved down a conveyor belt. Covariant's robot could pick and sort more than ten thousand different items with more than 99 percent accuracy. "I've worked in the logistics industry for more than sixteen years and I've never seen anything like this," said Peter Puchwein, vice president of Knapp, an Austrian company that had long provided automation technology for warehouses and helped develop and install the Covariant technology in Berlin. It showed that robotic automation would continue to spread across the retail and logistics industry in the years to come, and perhaps across manufacturing plants, too. It also raised new concerns over warehouse workers, losing their jobs to automated systems. In the German warehouse, the jobs of three humans were done by one robot.

At the time, though, economists didn't think this kind of technology would diminish the overall number of logistics jobs anytime soon. The online retail business was growing much too quickly, and most companies would take years or even decades to install the new breed of automation. But Abbeel acknowledged that at some point in the distant future, the situation would flip. He was also optimistic about the final outcome for humans. "If this happens fifty years from now," he said, "there is plenty of time for the educational system to catch up."

20

RELIGION

"MY GOAL IS TO SUCCESSFULLY CREATE BROADLY BENEFICIAL AGI. I ALSO UNDERSTAND THIS SOUNDS RIDICULOUS."

In the fall of 2016, three days before the premiere of *Westworld*—the HBO television series in which amusement park androids turn on their creators after slowly crossing the threshold into artificial sentience—many of the cast and crew attended a private screening in Silicon Valley. It was not held at the local cineplex. It was held at the home of Yuri Milner, a fifty-four-year-old Israeli-Russian entrepreneur and venture capitalist who was an investor in Facebook, Twitter, Spotify, and Airbnb and a regular at the Edge Foundation's annual Billionaires' Dinner. A twenty-five-thousand-five-hundred-square-foot limestone mansion perched in the Los Altos hills overlooking San Francisco Bay, his home was called Chateau Loire. Purchased five years earlier for more than $100 million, it was one of the most expensive single-family homes in the country, spanning indoor and outdoor swimming pools, a ballroom, a tennis court, a wine cellar, a library, a game room, a spa, a gym, and its own private movie theater.

When guests arrived for the screening, they were met at the gate by valets carrying iPads. The valets checked their invitations, marked their names on the iPads, parked their cars, shuttled them up the hill in golf carts, and dropped them in front of the movie theater, a freestanding building at the foot of the faux château. A red carpet led them to the door. Sergey Brin was among those who walked the red carpet, wearing what seemed to be a Native American rug draped over his shoulders like a shawl. Many of the other guests were the millionaire founders of start-ups that had recently emerged from Y Combinator, the start-up accelerator overseen by Sam Altman. Some were among the founders who'd responded to a mysterious invitation five years earlier, filed into a conference room inside the Y Combinator offices in San Francisco, and looked on in surprise as a robot rolled into the room with an iPad where its head should have been, a live closeup of Yuri Milner appeared on the iPad, and Milner suddenly announced he was investing $150,000 in each and every one of their brand-new companies.

Yuri Milner hosted the *Westworld* screening alongside Sam Altman. "Sam Altman + Yuri Milner invite you to a pre-release screening of the opening episode of *Westworld*, a new HBO series exploring the dawn of artificial consciousness and intelligence," the invitation read. After the screening, the show's cast and crew, including creator and director Jonathan Nolan and stars Evan Rachel Wood and Thandie Newton, walked onstage and sat atop high stools lined up in front of the movie screen. They spent the next hour discussing the episode, in which several of the Westworld androids begin to malfunction and misbehave after a software update gives them access to past memories. Then Altman took the stage alongside Ed Boyden, an MIT professor who specialized in nascent technologies for sending information between machines and the human brain. Boyden had recently won the Breakthrough Prize, a $3 million research grant created by Yuri Milner, Sergey Brin, Mark Zuckerberg, and other Silicon Valley luminaries. Alongside Altman, he told their private audience that scientists were approaching the point where they could create a complete map of the brain and then simulate it with a machine. The question was whether the machine, in addition to acting like a human, would actually *feel* what it was like to be human. This, they said, was the same question explored in *Westworld*.

AFTER Marvin Minsky, John McCarthy, and the other founding fathers of the AI movement gathered at Dartmouth in the summer of 1956, some said a machine would be intelligent enough to beat the world chess champion or prove its own mathematical theorem within a decade. A decade later, that hadn't happened. One of the founding fathers, Carnegie Mellon professor Herbert Simon, then said the field would deliver machines capable of "doing any work a man can do" within the next twenty years. But almost immediately, the first AI winter set in. When the thaw arrived in the 1980s, others—including Doug Lenat, who set out to rebuild common sense through the project he called Cyc—vowed to recreate human intelligence. But by the '90s, when Cyc showed little sign of real progress, the idea of rebuilding human intelligence was not something the leading researchers talked about, at least not in public, and that remained true for the next two decades. In 2008, Shane Legg said as much in his PhD thesis. "Among researchers the topic is almost taboo: it belongs in science fiction. The most intelligent computer in the world, they assure the public, is perhaps as smart as an ant, and that's on a good day. True machine intelligence, if it is ever developed, lies in the distant future," he wrote. "Perhaps over the next few years these ideas will become more mainstream, however for now they are at the fringe. Most researchers remain very sceptical about the idea of truly intelligent machines within their lifetime."

Over the next few years, such ideas did become more mainstream, and this was largely thanks to Shane Legg, who went on to build DeepMind alongside Demis Hassabis, and who, with Hassabis, convinced three significant figures (Peter Thiel, Elon Musk, and Larry Page) that the research was worth investing in. After Google acquired DeepMind, Legg continued to argue in private that superintelligence was nigh, but he rarely spoke about it in public, in part because people like Musk were intent on stoking concerns that intelligent machines could destroy the world. But in spite of his reticence, his ideas continued to spread.

When Ilya Sutskever had interviewed with Hassabis and Legg while still a grad student at the University of Toronto and the two DeepMind founders said they were building artificial general intelligence, he thought they had lost touch with reality. But after his personal success with image recognition and machine translation at Google—and after he spent several weeks inside DeepMind—he came to embrace Legg's thesis as "insanely visionary." Many others did, too. Five of OpenAI's first nine researchers had spent time inside the London lab where the possibilities of AGI had been so fervently embraced, and the two labs shared two investors: Thiel and Musk. In the fall of 2015, as Sutskever discussed the lab that would become OpenAI, he felt he'd found a group of people who thought as he did—who held the same beliefs and ambitions—but he worried their conversations would come back to haunt him. If others heard he was discussing the rise of artificial general intelligence, he would be branded a pariah across the wider community of researchers. When OpenAI was unveiled, the official announcement did not mention AGI. It only hinted at the idea as a distant possibility. "AI systems today have impressive but narrow capabilities," the announcement read. "It seems that we'll keep whittling away at their constraints, and in the extreme case, they will reach human performance on virtually every intellectual task." But as the lab grew, Sutskever shed his fears. When OpenAI hired Ian Goodfellow in 2016, a year after its launch, the lab welcomed him into the fold with drinks at a San Francisco bar, and Sutskever gave the toast, raising a glass high. "To AGI in three years!" he said. As he did, Goodfellow wondered if it was too late to tell the lab he didn't want the job after all.

Belief in AGI required a leap of faith. But it drove some researchers forward in a very real way. It was something like a religion. "As scientists, we often feel the need to justify our work in very pragmatic terms. We want to explain to people why what we're doing is valuable today. But often, what really drives scientists to do what they do is larger," says the roboticist Sergey Levine. "The thing that drives them is more emotional. It's more visceral than fundamental. That's why people are into AGI. They are a larger group than it might seem." As Alex Krizhevsky puts it: "We believe what we are emotionally inclined to believe in."

Belief in AGI had a way of spreading from person to person. Some were afraid to believe until enough people around them believed. And yet no one believed in quite the same way as everyone else. Each viewed the technology and its future through their own lens. Then the belief moved into Silicon Valley, and it was amplified. The Valley imbued this idea with more money, more showmanship, and more belief. Although researchers like Sutskever were initially reticent about voicing their views, Elon Musk did not hold back. Nor did the lab's other chairman: Sam Altman.

In the first days of 2017, the Future of Life Institute held another summit, this one in a tiny town on the central coast of California called Pacific Grove. Pacific Grove was home to Asilomar, the sprawling rustic hotel among the evergreens where the world's leading geneticists gathered in the winter of 1975 to discuss whether their work in gene editing would end up destroying the world. Now AI researchers gathered in the same beachside grove to discuss, once again, whether AI posed the same existential risk. Altman was there. So were Musk and most of the other big players from OpenAI and DeepMind. On the second day of the retreat, Musk took the stage as part of a nine-person panel dedicated to the idea of superintelligence. Each panelist was asked if it was possible, and as they passed the microphone down the line, each said "Yes," until the microphone reached Musk. "No," he said, as laughter cascaded across the small auditorium. They all knew what he believed. "We are headed toward either superintelligence or civilization ending," he said after the laughter died down. As the panel continued, Max Tegmark asked how humans could live alongside superintelligence once it arrived, and Musk said this would require a direct connection between brains and machines. "All of us already are cyborgs," he said. "You have an extension of yourself in your phone and your computer and all your applications. You are already superhuman." The limitation, he explained, was that people couldn't use their applications fast enough. There was not enough "bandwidth" between brain and machine. People still used "meat sticks"—fingers—to type stuff into their phones. "We have to solve that constraint with a high-bandwidth connection to the neural cortex."

At one point, Oren Etzioni, the head of the Allen Institute for Artificial Intelligence, took the stage and tried to temper this talk. "I hear a lot of people saying a lot of things without hard grounding in data," he said. "I encourage people to ask: 'Is this based on data or is this based on hardcore speculation?'" But others in the room sided with Musk. It was an argument that grew increasingly common at events across the community—and it was not an argument anyone could win. It was an argument about what would happen in the future, and that meant anyone could claim anything without being proven wrong. But more than most, Musk knew how to use this to his advantage. A few months later, he unveiled a new start-up, called Neuralink, backed by $100 million, that aimed to create a "neural lace"—an interface between computer and brain—and it moved into the same offices as OpenAI.

Though Musk soon left OpenAI, the lab's ambitions only grew under Altman. Sam Altman was a Silicon Valley archetype: In 2005, he founded a social networking company as a twenty-year-old college sophomore. The company was called Loopt, and it eventually raised $30 million in venture capital, including one of the first investments made by Y Combinator and its founder Paul Graham. Seven years later, Loopt's social networking service was shut down after being sold at a loss for its investors. But this was a successful exit for Altman, a trim, compact man with sharp green eyes and a particular talent for raising money. Graham soon announced that he was stepping down as president of Y Combinator, and he named Altman as his replacement, an appointment that surprised many across the family of Y Combinator companies. This made Altman an advisor to an endless stream of start-ups. In exchange for advice and capital, Y Combinator received a stake in each company, and Altman personally invested in some companies, too, becoming very wealthy, very quickly. He felt that any monkey could run YC, but he also felt that in running it, he developed an usually sharp talent for evaluating people, not to mention the skill and opportunity needed to raise large amounts of capital. During his rapid ascent, he was motivated first by money, then by the power over the people and the companies in his orbit, and then by the satisfaction that came from building companies that had a real impact on the larger world. With OpenAI, he aimed to make a much larger impact. The quest for AGI was more important—and more interesting—than anything else he could chase. Leaving Y Combinator for OpenAI, he believed, was the inescapable path.

Like Musk, he was an entrepreneur, not a scientist, though he sometimes made a point of saying he studied AI at Stanford before dropping out as a sophomore. Unlike Musk, he was not in constant search of attention and controversy in the press and on social media, but he, too, was someone who lived as if the future had already arrived. This was the norm among the Silicon Valley elite, who knew, either consciously or unconsciously, that it was the best way of attracting attention, funding, and talent, whether they were inside a large company or launching a small start-up. Ideas might fail. Predictions might not be met. But the next idea wouldn't succeed unless they, and everyone around them, believed that it could. "Self-belief is immensely powerful. The most successful people I know believe in themselves almost to the point of delusion," he once wrote. "If you don't believe in yourself, it's hard to let yourself have contrarian ideas about the future. But this is where most value gets created." He then recalled the time Musk took him on a tour of the SpaceX factory, when he was struck not so much by the rockets designed for a trip to Mars but by the look of certainty on Musk's face. "Huh," Altman thought to himself, "so that's the benchmark for what conviction looks like."

Altman knew that what he believed wouldn't always come to pass. But he also knew that most people underestimated what time and rapid expansion could bring to seemingly small ideas. In the Valley, this was called "scale." When Altman decided an idea would scale, he was not afraid to bet big on its progress. He might be wrong time and again, but when he was right, he wanted to be breathtakingly right. For him, this attitude was encapsulated by an oft-repeated quote from Machiavelli: "Make mistakes of ambition and not mistakes of sloth." He lamented that in the wake of the 2016 election, the public did not champion the goals of Silicon Valley in the same way they'd backed, say, the Apollo program in the '60s—they viewed its ambitions not as something inspiring or cool but as self-indulgent or even harmful.

As OpenAI was announced, Altman wasn't as shy as Sutskever about the idea of rebuilding intelligence. "As time rolls on and we get closer to something that surpasses human intelligence, there is some question how much Google will share," he said. When he was asked if OpenAI would build the same technology, he said he expected it would, but he also said OpenAI would share what it built. "It will just be open source and usable by everyone instead of usable by, say, just Google." Artificial intelligence was a bigger idea than any other Altman had embraced, but he came to see it much as he saw the rest.

In April of 2018, he and his researchers released a new charter for the lab, describing a very different mission from the one he laid out as it was founded. Altman had originally said OpenAI would openly share all its research. That's why it was called OpenAI. But after seeing the turmoil created by the rise of generative models and face recognition and the threat of autonomous weapons, he now said that as time went on, it would hold back some technologies as it gauged their effect on the world at large. This was a reality that many organizations were now waking up to. "If you decide from the outset that it's an open platform that anyone can use however they like, the consequences are going to be significant," Mustafa Suleyman says. "One has to think much more sensitively about how technology is going to be misused before it is created, and how you can put a process in place that creates some oversight." The irony was that OpenAI took this attitude to extremes. In the months that followed, it became a new way for the lab to market itself. After building a new language model along the lines of Google BERT, OpenAI made a point of saying, through the press, that the technology was too dangerous to release because it would allow machines to automatically generate fake news and other misleading information. Outside the lab, many researchers scoffed at the claim, saying the technology wasn't even close to dangerous. And it was, eventually, released.

At the same time, the new OpenAI charter said—explicitly and matter-of-factly—that the lab was building AGI. Altman and Sutskever had seen both the limitations and the dangers of the current technologies, but their goal was a machine that could do anything the human brain could do. "OpenAI's mission is to ensure that artificial general intelligence (AGI)—by which we mean highly autonomous systems that outperform humans at most economically valuable work—benefits all of humanity. We will attempt to directly build safe and beneficial AGI, but will also consider our mission fulfilled if our work aids others to achieve this outcome," the charter read. What both Altman and Sutskever now said was that they would build general intelligence in much the same way DeepMind built systems that mastered Go and the other games. They said it was merely a matter of gathering enough data and building enough computing power and improving the algorithms that analyzed the data. They knew others were skeptical, and they believed the technology could be dangerous. But none of that bothered them. "My goal is to successfully create broadly beneficial AGI," Altman says. "I also understand this sounds ridiculous."

Later that year, DeepMind trained a machine to play capture the flag. This is a team sport played by children at summer camps, in the woods or across open fields, but it was also played by professional video gamers inside three-dimensional games like Overwatch and Quake III. The DeepMind researchers trained their machine inside Quake III, where the two flags, one red and one blue, sat at opposite ends of a high-walled maze. Each team guarded its own flag while also trying to capture their opponent's flag and bring it back to home base. It is a game that requires teamwork—a careful coordination between defense and attack—and the DeepMind researchers showed that machines could learn this kind of collaborative behavior, or at least learn to mimic it. Their system learned by playing about four hundred and fifty thousand rounds of Quake III capture the flag—more than four years of game play packed into a few weeks of training. In the end, it could play the game alongside other autonomous systems or with human players, tailoring its behavior to each teammate. In some cases, it exhibited the same collaborative skills as any other seasoned player. When a teammate was on the verge of capturing the flag, it would race to the opponent's home base. As human players knew, once a flag was captured, another would appear at the opponent's base, and as soon as it appeared, it could be taken. "How you define teamwork is not something I want to tackle," said Max Jaderberg, one of the DeepMind researchers who worked on the project. "But one agent will sit in the opponent's base camp, waiting for the flag to appear, and that is only possible if it is relying on its teammates."

This was how both DeepMind and OpenAI hoped to mimic human intelligence. Autonomous systems would learn inside increasingly complex environments. First Atari. Then Go. Then three-dimensional multiplayer games like Quake III that involved not just individual skills but also teamwork. And so on and so forth. Seven months later, DeepMind unveiled a system that beat the world's top professionals at StarCraft, a three-dimensional game set in space. Then OpenAI built a system that mastered Dota 2, a game that plays like a more complex version of capture the flag, requiring collaboration between entire teams of autonomous agents. That spring, a team of five autonomous agents beat a team of the world's best human players. The belief was that success in the virtual arena would eventually lead to automated systems that could master the real world. This was what OpenAI did with its robotic hand, training a virtual re-creation of the hand to solve a virtual Rubik's Cube before moving this know-how into the real world. If they could build a large enough simulation of what humans encounter in their daily lives, these labs believed, they could build AGI.

Others saw this work differently. As impressive as these feats were with Quake and StarCraft and Dota, many questioned how well they would translate to the real world. "3-D environments are designed to make navigation easy," Georgia Tech professor Mark Riedl said when DeepMind published a paper describing its capture-the-flag–playing agents. "Strategy and coordination in Quake are simple." Though these agents seemed to be collaborating, he said, they were not. They were merely responding to what was happening in the game, rather than actually communicating with one another, as human players might. Each had a superhuman knowledge of the game, but they were in no way intelligent. That meant they would struggle in the real world.

Reinforcement learning was ideally suited to games. Video games tallied points. But in the real world, no one was keeping score. Researchers had to define success in other ways, and that was far from trivial. A Rubik's Cube was very real, but it was also a game. The goal was easily defined. And still, it wasn't a problem that was completely solved. In the real world, OpenAI's robotic hand was equipped with tiny LEDs that allowed sensors elsewhere in the room to track exactly where each finger was at any given moment. Without these LEDs and sensors, it could not solve the Cube. And even with them, as the fine print said in OpenAI's research paper, it dropped the Cube eight times out of ten. To reach this 20 percent success rate, OpenAI's robotic hand drew on ten thousand years of digital trial and error. True intelligence would require a level of digital experience that would make this look minuscule. DeepMind could draw on Google's network of data centers, one of the largest private networks on Earth, but that was not enough.

The hope was that researchers could change the equation with new kinds of computer chips—chips that could drive this research to levels beyond both Nvidia's GPUs and Google's TPUs. Dozens of companies, including Google, Nvidia, and Intel, as well as a long line of start-ups, were building new chips just for training neural networks, so that systems built by labs like DeepMind and OpenAI could learn far more in far less time. "I look at what is coming in terms of new computing resources, and I plot this in relation to current results, and the curve keeps going up," Altman says.

With an eye on this new breed of hardware, Altman struck a deal with Microsoft and its new CEO, Satya Nadella, who were still trying to show the world that they remained a leader in artificial intelligence. In just a few short years, Nadella had turned the company around, embracing open-source software and racing ahead of Google in the cloud computing market. But in a world where many believed that the future of the cloud computing market was artificial intelligence, few saw Microsoft as a top player in the field. Nadella and company agreed to invest $1 billion in OpenAI, and OpenAI agreed to send much of this money back into Microsoft as the tech giant built a new hardware infrastructure just for training the lab's systems. "Whether it's our pursuit of quantum computing or it's a pursuit of AGI, I think you need these high-ambition North Stars," Nadella said. For Altman, this was less about the means than the end. "My goal in running OpenAI is to successfully create broadly beneficial AGI," he said. "This partnership is the most important milestone so far on that path."

Two labs now said they were building AGI. And two of the world's largest companies said they would provide the money and the hardware they would need along the way, at least for a while. Altman believed he and OpenAI would need another $25 billion to $50 billion to reach their goal.

ONE afternoon, Ilya Sutskever sat down in a coffee shop a few blocks from the OpenAI offices in San Francisco. As he sipped from a ceramic mug, he talked about many things, one of them AGI. He described it as a technology that he knew was coming even if he couldn't explain the particulars. "I know it's going to be huge. I think that with certainty," he said. "It's very hard to articulate exactly what it will look like, but I think it's important to think about these questions and see ahead as much as possible." He said it would be a "tsunami of computing," an avalanche of artificial intelligence. "This is almost like a natural phenomenon," he explained. "It's an unstoppable force. It's too useful to not exist. What can we do? We can steer it, move it this way or that way."

This was not something that would merely change the digital world. It would change the physical, too. "I think one could make a pretty good case that truly human-level AI and beyond will have an overwhelming and transformational impact on society in ways that are hard to predict and imagine," he said. "I think it will deconstruct pretty much all human systems. I think that it's fairly likely that it will not take too long of a time for the entire surface of the Earth to become covered with data centers and power stations. Once you have one data center, which runs lots of AIs on it, which are much smarter than humans, it's a very useful object. It can generate a lot of value. The first thing you ask it is: Can you please go and build another one?"

Asked if he meant this literally, he said he did, pointing out the window toward the bright orange building across the street from the coffee shop. Imagine, he said, that the building was filled with computer chips, and that these chips were running software that duplicated the skills of the chief executive officer of a company like Google as well as the chief financial officer and all its engineers. If you had all of Google running inside this building, he explained, it would be enormously valuable. It would be so valuable, it would want to erect another building just like itself. And another. And another. There would be enormous pressure, he said, to keep building more.

Across the Atlantic, inside the new Google building near St. Pancras station, Shane Legg and Demis Hassabis described the future in simpler terms. But their message wasn't all that different. As Legg explained it, DeepMind was on the same trajectory that he and Hassabis had envisioned when they first pitched the company to Peter Thiel ten years earlier. "When I look back at what we wrote about our mission at the start of the company, it feels incredibly similar to DeepMind today," he says. "It hasn't really changed at all." Just recently, they had shed the one part of their operation that didn't seem to fit that mission. As far back as the spring of 2018, Mustafa Suleyman told some at DeepMind that he would soon move the lab's health practice to Google, and that fall, DeepMind announced that Google was taking over the practice. A year later, after Suleyman took a leave of absence that was not initially revealed to anyone outside the company, he, too, left DeepMind for Google. His philosophy had always seemed more aligned with Jeff

Dean's than with Demis Hassabis's, and now he and his pet project, the most practical and near-term part of DeepMind, had parted ways with Hassabis and Legg. More than ever, DeepMind was focused on the future. And though it operated with considerable independence, it could still draw on Google's vast resources. Since acquiring DeepMind, Google had invested $1.2 billion in its research. By 2020, in addition to the hundreds of computer scientists at the London lab, Hassabis had hired a team of more than fifty neuroscientists to investigate the inner workings of the brain.

Some questioned how long this would last. The same year, Larry Page and Sergey Brin, DeepMind's biggest supporters, announced they were retiring. "Would DeepMind continue to receive such large amounts of Alphabet money for such long-term research?" the voices asked. "Or would it be forced onto more immediate tasks?" For Alan Eustace, the man who led the acquisition of DeepMind and helped build Google Brain, there would always be tension between the chase for near-term technology and the distant dream. "It could be that inside Google, they would have access to more interesting problems, but it might slow down their long-term goal. Putting them in Alphabet slows their ability to commercialize their technology, but it is more likely to have a positive long-term effect," he says. "The solution to this conundrum is an important step in the history of machine learning." But certainly, the philosophy that drove DeepMind hadn't changed. After years of turmoil, when AI technologies improved at such a surprising rate and behaved in ways no one expected and intertwined with corporate forces more powerful and more relentless than anyone realized, DeepMind, like OpenAI, was still intent on building a truly intelligent machine. In fact, its founders saw the turmoil as a kind of vindication. They had warned that these technologies could go wrong.

One afternoon, during a video call from his office in London, Hassabis said his views came down somewhere between Mark Zuckerberg's and Elon Musk's. The views of both Zuckerberg and Musk, he said, were extreme. He very much believed that superintelligence was possible and he believed it could be dangerous, but he also believed it was still many years away. "We need to use the downtime, when things are calm, to prepare for when things get serious in the decades to come," he said. "The time we have now is valuable, and we need to make use of it." The problems brought on by Facebook and other companies in recent years were a warning that these technologies must be built in careful and thoughtful ways. But this warning would not stop him from reaching his goal. "We're doing this," he said. "We're not messing around. We're doing this because we really believe it's possible. The timescales are debatable, but as far as we know, there's no law of physics that prevents AGI being built."

21

X FACTOR

"HISTORY IS GOING TO REPEAT ITSELF,
I THINK."

Inside Geoff Hinton's office on the fifteenth floor of the Google building in downtown Toronto, two white blocks sat on the cabinet near the window. Each was about the size of a shoebox. Oblong, sharp-edged, and triangular, they looked like two matching modernist mini-sculptures he'd found at the back of an IKEA catalog. When someone new walked into his office, he would hand them these two blocks, explain that they were now holding two halves of the same pyramid, and ask if they could put the pyramid back together. This seemed like a simple task. Each block had only five faces, and all anyone had to do was find the two faces that matched and line them up. But few could solve the puzzle. Hinton liked to say that two tenured MIT professors failed to solve it. One refused to try. The other produced a proof that it wasn't possible.

But it was possible, Hinton said, before quickly solving the puzzle himself. Most people failed the test, he explained, because the puzzle undermined the way they understood an object like a pyramid—or anything else they might encounter in the physical world. They didn't recognize a pyramid by looking at one side and then the other and then the top and then the bottom. They pictured the whole thing *sitting in three-dimensional space.* Because of the way his puzzle cut the pyramid in two, Hinton explained, it prevented people from picturing it in three dimensions as they normally would. This was his way of showing that vision was more complex than it might seem, that people understood what was in front of them in ways machines still could not. "It is a fact that is ignored by researchers in computer vision," he said. "And that is a huge mistake."

He was pointing to the limitations of the technology he helped build over the last four decades. Researchers in computer vision now relied on deep learning, he said, and deep learning solved only part of the problem. If a neural network analyzed thousands of coffee cup photos, it could learn to recognize a coffee cup. But if those photos pictured coffee cups only from the side, it couldn't recognize a coffee cup turned upside down. It saw objects in only two dimensions, not three. This, he explained, was one of the many problems he hoped to solve with "capsule networks," using a very English pronunciation: cap-*shule.*

Like any other neural network, a capsule network is a mathematical system that learned from data. But, Hinton said, it could give machines the same three-dimensional perspective as humans, letting them recognize a coffee cup from any angle after learning what it looked like from only one. It was an idea he first developed in the late 1970s, before reviving it at Google decades later. When he spent the summer at DeepMind in 2015, this was what he hoped to work on but couldn't after his wife, Jackie, was diagnosed with cancer. Once back in Toronto, he explored the idea alongside Sara Sabour, the Iranian researcher who had been denied a U.S. visa, and by the fall of 2017, they'd built a capsule network that could recognize images from unfamiliar angles with an accuracy beyond what an ordinary neural network could do. But capsule networks, he explained, were not just a way of recognizing images. They were an attempt to mimic the brain's network of neurons in a far more complex and powerful way, and he believed they could accelerate the progress of artificial intelligence as a whole, from computer vision to natural language understanding and beyond. For Hinton, this new technology was approaching a point much like the one that neural networks reached in December 2008 when he ran into Li Deng at the ski resort in Whistler. "History is going to repeat itself," he said, "I think."

ON March 27, 2019, the Association for Computing Machinery, the world's largest society of computer scientists, announced that Hinton, LeCun, and Bengio had won the Turing Award. First introduced in 1966, the Turing Award was often called "the Nobel Prize of computing." It was named for Alan Turing, one of the key figures in the creation of the computer, and it now came with $1 million in prize money. After reviving neural network research in the mid-2000s and pushing it into the heart of the tech industry, where it remade everything from image recognition to machine translation to robotics, the three veteran researchers split the prize three ways, with LeCun and Bengio giving the extra cent to Hinton.

Hinton marked the occasion with a rare tweet that described what he called the "X Factor." "When I was an undergrad at King's College Cambridge, Les Valiant, who won the Turing Award in 2010, lived in the adjacent room on X staircase," the tweet read.

"He just told me that Turing lived on X staircase when he was a fellow at King's and probably wrote his 1936 paper there." This was the paper that helped launch the computer age.

The awards ceremony was held two months later in the Grand Ballroom of the Palace Hotel in downtown San Francisco. Jeff Dean attended in black tie. So did Mike Schroepfer. A waitstaff in white jackets served dinner to more than five hundred guests sitting at round tables with white tablecloths, and as they ate, various other awards were presented to more than a dozen engineers, programmers, and researchers from across industry and academia. Hinton did not sit and eat. Thanks to his bad back, fifteen years had now passed since he'd last sat down. "It's been a long-standing problem," he often said. As the first awards were given, he stood at the side of the ballroom, looking down at a small card where he'd written the notes for his speech. For a while, LeCun and Bengio stood beside him near the wall. Then they sat down with the rest as Hinton continued to read his card.

After the first hour of awards, Jeff Dean walked up onto the stage and rather nervously introduced the three Turing winners. He was a world-class engineer, not an orator. But his words were true. Despite years of doubt from the rest of the field, he told the room, Hinton, LeCun, and Bengio had developed a set of technologies that were still shifting both the scientific and the cultural landscape. "It is time to recognize the greatness of work with a contrarian point of view," he said. Then a brief video played across two screens on either side of the stage, describing the long history of neural networks and the resistance these three researchers had faced over the decades. When it cut to Yann LeCun, he said: "I was definitely thinking I was right the whole time." As the laughter rippled across the ballroom, Hinton, now standing onstage, kept reading his card.

The video was careful to show that artificial intelligence was still a long way from true intelligence. Up on the screen, LeCun said: "Machines still have less common sense than a house cat." Then the video cut to Hinton describing his work with capsules, which he hoped would push the field forward once again, and a voice-over narration delivered the usual hyperbole. "The future of AI looks promising," it said, "in what many are calling the Next Big Thing with endless possibilities." Then Hinton appeared on-screen one last time, describing the moment in simpler terms. First, he said he was pleased to win the award alongside LeCun and Bengio. "It is very nice to win this as a group," he said. "It is always nicer to be part of a successful group than to be on your own." Then he gave the room some advice. "If you have an idea and it seems to you it has to be right, don't let people tell you it's silly," he said. "Just ignore them." Standing onstage as the video ended, he still looked down at his card.

Bengio gave the first acceptance speech, now wearing a full beard that was mostly gray. He was first, he said, because he was the youngest of the three. He thanked the Canadian Institute for Advanced Research, the government organization that funded their neural network research in the mid-2000s, and he thanked LeCun and Hinton. "They were first my role models and then my mentors and then my friends and partners in crime," he said. This was a prize not just for the three of them, he added, but for all the other researchers who believed in these ideas, including their many students at the University of Montreal, NYU, and the University of Toronto. What would ultimately push the technology to new heights, he said, was new research from the much wider community of like-minded thinkers. As he spoke, this was already happening, with progress continuing in healthcare, robotics, and natural language understanding. Over the years, the influence of many of the field's biggest names had waxed and waned. After growing disillusioned with his work, Alex Krizhevsky resigned from Google and left the field entirely. The next year, after increased tension at the highest rungs of the Baidu hierarchy, first Andrew Ng and then Qi Lu left the Chinese company. But the field as a whole continued to expand across both industry and academia. Over the last several months, Apple had poached both John Giannandrea and Ian Goodfellow, Bengio's former student, from Google.

But Bengio also warned that the community must be mindful of how these technologies were used. "The honor we are getting comes with responsibility," he said. "Our tools can be used in good and bad ways." Two months earlier, the *New York Times* had revealed that the Chinese government, in tandem with various AI companies, had built face recognition technology that could help track and control Uighurs, a largely Muslim ethnic minority. The following fall, Google general counsel Kent Walker said that despite the controversy over Project Maven, it was willing to work with the Department of Defense, and many still questioned what this kind of work meant for the future of autonomous weapons. The 2020 election was looming.

LeCun, the head of the AI lab at Facebook, was the next to speak. "Following Yoshua is always a challenge," he said. "Preceding Geoff even more." The only one of the three wearing a tuxedo, he said that many people had asked him how winning the Turing Award had changed his life. "Before, I was getting used to people telling me I was wrong," he said. "Now, I have to be careful because no one will dare tell me that I am wrong." He said that he and Bengio were unique among Turing Award winners. They were the only two born as late as the 1960s. They were the only two born in France. They were the only two whose first names began with *Y*.

And they were the only two with brothers who worked for Google. He thanked his father for teaching him to be an engineer, and he thanked Geoff Hinton for being a mentor.

As the room applauded LeCun, Hinton put away his card and walked to the lectern. "I have been doing some math," he said. "I am fairly sure I am younger than Yann plus Yoshua." He thanked "the ACM awards committee and their extraordinary good sense." He thanked his students and postdocs. He thanked his mentors and colleagues. He thanked the organizations who funded his research. But the person he really wanted to thank, he said, was his wife, Jackie. She had died a few months before the award was announced. Twenty-five years earlier, he told the room, his wife, Rosalind, had died and he had thought his research career was over. "A few years later, Jackie abandoned her own career in London and moved to Canada with the rest of us," he said, as his voice broke. "Jackie knew how much I wanted this award and she would have wanted to be here today."

DURING his Turing lecture in Phoenix, Arizona, the following month—given in celebration of the award—Hinton explained the rise of machine learning and explored where it might be going. At one point, he described the various ways a machine could learn. "There are two kinds of learning algorithms—actually three, but the third doesn't work very well," he said. "That is called reinforcement learning." His audience, several hundred AI researchers strong, let out a laugh. So he went further. "There is a wonderful *reductio ad absurdum* of reinforcement learning," he told the crowd. "It is called DeepMind." Hinton did not believe in reinforcement learning, the method Demis Hassabis and DeepMind saw as the path to AGI. It required too much data and too much processing power to succeed with practical tasks in the real world. For many of the same reasons—and many others—he didn't believe in the race to AGI, either.

AGI, he believed, was a task far too big to be solved in the foreseeable future. "I'd much rather focus on something where you can figure out how you might solve it," he said during a visit to Google headquarters in Northern California that spring. But he also wondered why anyone would want to build it. "If I have a robot surgeon, it needs to understand an awful lot about medicine and about manipulating things. I don't see why my robot surgeon needs to know about baseball scores. Why would it need general-purpose knowledge? I would have thought you would make your machines to help us," he said. "If I want a machine to dig a ditch right, I'd rather have a backhoe than an android. You don't want an android digging a ditch. If I want a machine to dispense some money, I want an ATM. One thing I believe is that we probably don't want general-purpose androids." When asked if belief in AGI was something like a religion, he demurred: "It's not nearly as dark as religion."

That same year, Pieter Abbeel asked him to invest in Covariant. And when Hinton saw what reinforcement learning could do for Abbeel's robots, he changed his mind about the future of AI research. As Covariant's systems moved into the warehouse in Berlin, he called it "the AlphaGo moment" for robotics. "I have always been skeptical of reinforcement learning, because it required an extraordinary amount of computation. But we've now got that," he said. Still, he didn't believe in building AGI. "The progress is being made by tackling individual problems—getting a robot to fix things or understanding a sentence so that you can translate—rather than people building general AI," he said.

At the same time, he didn't see an end to the progress across the field, and it was now out of his hands. He hoped for one last success with capsules, but the larger community, backed by the world's biggest companies, was racing in other directions. Asked if we should worry about the threat of superintelligence, he said this didn't make much sense in the near term. "I think we are much better off than Demis thinks," he said. But he also said it was a perfectly reasonable worry if you looked into the distant future.

ACKNOWLEDGMENTS

This was not the book I was going to write. By the summer of 2016, I had spent several months writing a proposal for a very different book on a very different subject when I was contacted by a literary agent who was not the one I was already working with. This was Ethan Bassoff, and after he read the proposal I had written, he told me, very politely, that it was garbage. He was right. More important, he believed in the idea that became this book—an idea no one else really believed in. Those are always the best ideas.

Through Ethan, I found my editor at Dutton, Stephen Morrow, and he believed in this idea, too. That is remarkable. I was pitching a book about artificial intelligence, which happened to be the most hyped technology on Earth, but my idea was to write a book not about the technology but about the people building it. No one really writes books about the people who build technology. They write books about the executives who run the companies that build the technology. I was lucky to meet both Ethan and Stephen. And I was lucky that the people I wanted to write about were so interesting and so eloquent and so completely different from one other.

Then I had to actually write the book. This happened thanks almost entirely to my wife, Tay, and my daughters, Millay and Hazel. There must have been many times when they thought that trying to fit this book into everything else we had to do was not the wisest decision. If they did, they were right. But they helped me write it anyway.

Two people taught me how to write this book, just by example: Ashlee Vance and Bob McMillan. I didn't really know what to do with myself until I met Ashlee and our editor, Drew Cullen, at the *Register*, a truly strange and truly wonderful publication where I worked for five years. After that, Bob and I built what must surely be the greatest publication the world has ever seen: *Wired Enterprise.* It was another idea no one believed in, except for our boss, Evan Hansen. Evan: I owe you. I love doing stuff that seems like it should never work, and at *Wired*, I was able to do that for many years.

During those years, I wasn't the one who laid the groundwork for the "deep learning" beat. That was Bob and our favorite *Wired* science reporter, Daniela Hernandez. I am enormously grateful to Daniela and the rest of the kind few who agreed to read the first draft of my manuscript. That includes Oren Etzioni, a veteran AI researcher who was willing to provide an objective view while reading about his own field, and Chris Nicholson, who has been essential both to this book and to my coverage at the *New York Times*. He is always the rabbi I need.

I must also thank my editors Pui-Wing Tam and Jim Kerstetter, who have given me the job I always wanted and have taught me so much about how to do it well. And I must thank all my other colleagues at the *New York Times*, both in the San Francisco bureau and elsewhere, especially the great Scott Shane and the equally great Dai Wakabayashi, who worked with me on a story that helped open new doors for this book; Adam Satariano, Mike Isaac, Brian Chen, and Kate Conger, remarkably talented reporters who collaborated with me on other stories that play big roles here; and Nellie Bowles, who suggested what became the title of the prologue, which I am very pleased with.

Last, I send thanks to my mother, Mary Metz; my sisters, Louise Metz and Anna Metz Lutz; my brothers-in-law, Anil Gehi and Dan Lutz; and my nephews and nieces, Pascal Gehi, Elias Gehi, Miriam Lutz, Isaac Lutz, and Vivian Lutz, who have all helped much more than they know. My only regret is that I didn't finish this book soon enough for my father, Walt Metz, to read it. He would have enjoyed it more than anyone.

TIMELINE

1960—Cornell professor Frank Rosenblatt builds the Mark I Perceptron, an early "neural network," at a lab in Buffalo, New York.

1969—MIT professors Marvin Minsky and Seymour Papert publish *Perceptrons*, pinpointing the flaws in Rosenblatt's technology.

1971—Geoff Hinton starts a PhD in artificial intelligence at the University of Edinburgh.

1973—The first AI winter sets in.

1978—Geoff Hinton starts a postdoc at the University of California–San Diego.

1982—Carnegie Mellon University hires Geoff Hinton.

1984—Geoff Hinton and Yann LeCun meet in France.

1986—David Rumelhart, Geoff Hinton, and Richard Williams publish their paper on "backpropagation," expanding the powers of neural networks.

Yann LeCun joins Bell Labs in Holmdel, New Jersey, where he begins building LeNet, a neural network that can recognize handwritten digits.

1987—Geoff Hinton leaves Carnegie Mellon for the University of Toronto.

1989—Carnegie Mellon graduate student Dean Pomerleau builds ALVINN, a self-driving car based on a neural network.

1992—Yoshua Bengio meets Yann LeCun while doing postdoctoral research at Bell Labs.

1993—The University of Montreal hires Yoshua Bengio.

1998—Geoff Hinton founds the Gatsby Neuroscience Unit at University College London.

1990s–2000s—Another AI winter.

2000—Geoff Hinton returns to the University of Toronto.

2003—Yann LeCun moves to New York University.

2004—Geoff Hinton starts "neural computation and adaptive perception" workshops with funding from the Canadian government. Yann LeCun and Yoshua Bengio join him.

2007—Geoff Hinton coins the term "deep learning," a way of describing neural networks.

2008—Geoff Hinton runs into Microsoft researcher Li Deng in Whistler, British Columbia.

2009—Geoff Hinton visits Microsoft Research lab in Seattle to explore deep learning for speech recognition.

2010—Abdel-rahman Mohamed and George Dahl, two of Hinton's students, visit Microsoft.

Demis Hassabis, Shane Legg, and Mustafa Suleyman found DeepMind.

Stanford professor Andrew Ng pitches Project Marvin to Google chief executive Larry Page.

2011—University of Toronto researcher Navdeep Jaitly interns at Google in Montreal, building a new speech recognition system through deep learning.

Andrew Ng, Jeff Dean, and Greg Corrado found Google Brain.

Google deploys speech recognition service based on deep learning.

2012—Andrew Ng, Jeff Dean, and Greg Corrado publish the Cat Paper.

Andrew Ng leaves Google.

Geoff Hinton "interns" at Google Brain.

Geoff Hinton, Ilya Sutskever, and Alex Krizhevsky publish the AlexNet paper.

Geoff Hinton, Ilya Sutskever, and Alex Krizhevsky auction their company, DNNresearch.

2013—Geoff Hinton, Ilya Sutskever, and Alex Krizhevsky join Google.

Mark Zuckerberg and Yann LeCun found the Facebook Artificial Intelligence Research lab.

2014—Google acquires DeepMind.

Ian Goodfellow publishes the GAN paper, describing a way of generating photos.

Ilya Sutskever unveils the Sequence to Sequence paper, a step forward for automatic translation.

2015—Geoff Hinton spends the summer at DeepMind.

AlphaGo defeats Fan Hui in London.

Elon Musk, Sam Altman, Ilya Sutskever, and Greg Brockman found OpenAI.

2016—DeepMind unveils DeepMind Health.

AlphaGo defeats Lee Sedol in Seoul, South Korea.

Qi Lu leaves Microsoft.

Google deploys translation service based on deep learning.

Donald Trump defeats Hillary Clinton.

2017—Qi Lu joins Baidu.

AlphaGo defeats Ke Jie in China.

China unveils national AI initiative.

Geoff Hinton unveils capsule networks.

Nvidia unveils progressive GANs, which can generate photo-realistic faces.

Deepfakes arrive on the Internet.

2018—Elon Musk leaves OpenAI.

Google employees protest Project Maven.

Google releases BERT, a system that learns language skills.

2019—Top researchers protest Amazon face recognition technology.

Geoff Hinton, Yann LeCun, and Yoshua Bengio win the Turing Award for 2018.

Microsoft invests $1 billion in OpenAI.

2020—Covariant unveils "picking" robot in Berlin.

THE PLAYERS

AT GOOGLE

ANELIA ANGELOVA, the Bulgaria-born researcher who brought deep learning to the Google self-driving car project alongside Alex Krizhevsky.

SERGEY BRIN, founder.

GEORGE DAHL, the English professor's son who explored speech recognition alongside Hinton in Toronto and at Microsoft before joining Google Brain.

JEFF DEAN, the early Google employee who became the company's most famous and revered engineer before founding Google Brain, its central artificial intelligence lab, in 2011.

ALAN EUSTACE, the executive and engineer who oversaw Google's rush into deep learning before leaving the company to set a world skydiving record.

TIMNIT GEBRU, the former Stanford researcher who joined the Google ethics team.

JOHN "J.G." GIANNANDREA, the head of AI at Google who defected to Apple.

IAN GOODFELLOW, the inventor of GANs, a technology that could generate fake (and remarkably realistic) images on its own, who worked at both Google and OpenAI before moving to Apple.

VARUN GULSHAN, the virtual reality engineer who explored AI that could read eye scans and detect signs of diabetic blindness.

GEOFF HINTON, the University of Toronto professor and founding father of the "deep learning" movement who joined Google in 2013.

URS HÖLZLE, the Swiss-born engineer who oversaw Google's global network of computer data centers.

ALEX KRIZHEVSKY, the Hinton protégé who helped remake computer vision at the University of Toronto before joining Google Brain and the Google self-driving car project.

FEI-FEI LI, the Stanford professor who joined Google and pushed for a Google AI lab in China.

MEG MITCHELL, the researcher who left Microsoft for Google, founding a team dedicated to the ethics of artificial intelligence.

LARRY PAGE, founder.

LILY PENG, the trained physician who oversaw a team that applied artificial intelligence to healthcare.

SUNDAR PICHAI, CEO.

SARA SABOUR, the Iran-born researcher who worked on "capsule networks" alongside Geoff Hinton at the Google lab in Toronto.

ERIC SCHMIDT, chairman.

AT DEEPMIND

ALEX GRAVES, the Scottish researcher who built a system that could write in longhand.

DEMIS HASSABIS, the British chess prodigy, game designer, and neuroscientist who founded DeepMind, a London AI start-up that would grow into the world's most celebrated AI lab.

KORAY KAVUKCUOGLU, the Turkish researcher who oversaw the lab's software code.

SHANE LEGG, the New Zealander who founded DeepMind alongside Demis Hassabis, intent on building machines that could do anything the brain could do—even as he worried about the dangers this could bring.

VLAD MNIH, the Russian researcher who oversaw the creation of a machine that mastered old Atari games.

DAVID SILVER, the researcher who met Hassabis at Cambridge and led the DeepMind team that built AlphaGo, the machine that marked a turning point in the progress of AI.

MUSTAFA SULEYMAN, the childhood acquaintance of Demis Hassabis who helped launch DeepMind and led the lab's efforts in ethics and healthcare.

AT FACEBOOK

LUBOMIR BOURDEV, the computer vision researcher who helped create the Facebook lab.

ROB FERGUS, the researcher who worked alongside Yann LeCun at both NYU and Facebook.

YANN LECUN, the French-born NYU professor who helped nurture deep learning alongside Geoff Hinton before overseeing the Facebook Artificial Intelligence Research lab.

MARC'AURELIO RANZATO, the former professional violinist who Facebook poached from Google Brain to seed its AI lab.

MIKE "SCHREP" SCHROEPFER, chief technology officer.

MARK ZUCKERBERG, founder and CEO.

AT MICROSOFT

CHRIS BROCKETT, the former professor of linguistics who became a Microsoft AI researcher.

LI DENG, the researcher who brought Geoff Hinton's ideas to Microsoft.

PETER LEE, head of research.

SATYA NADELLA, CEO.

AT OPENAI

SAM ALTMAN, the president of Silicon Valley start-up incubator Y Combinator who became OpenAI's CEO.

GREG BROCKMAN, the former chief technology officer of fintech start-up Stripe who helped build OpenAI.

ELON MUSK, the CEO of electric car maker Tesla and rocket company SpaceX who helped create OpenAI.

ILYA SUTSKEVER, the Geoff Hinton protégé who left Google Brain to join OpenAI, the San Francisco AI lab created in response to DeepMind.

WOJCIECH ZAREMBA, the former Google and Facebook researcher who was one of OpenAI's first hires.

AT BAIDU

ROBIN LI, CEO.

QI LU, the Microsoft executive vice president who oversaw the Bing search engine before leaving the company and joining Baidu.

ANDREW NG, the Stanford University professor who founded the Google Brain lab alongside Jeff Dean before taking over Baidu's Silicon Valley lab.

KAI YU, the researcher who founded the deep learning lab at Chinese tech giant Baidu.

AT NVIDIA

CLÉMENT FARABET, the Yann LeCun protégé who joined Nvidia's efforts to build deep learning chips for driverless cars.

JENSEN HUANG, CEO.

AT CLARIFAI

DEBORAH RAJI, the Clarifai intern who went on to explore bias in AI systems at MIT.

MATTHEW ZEILER, founder and CEO.

IN ACADEMIA

YOSHUA BENGIO, the University of Montreal professor who carried the torch for deep learning alongside Geoff Hinton and Yann LeCun in the 1990s and 2000s.

JOY BUOLAMWINI, the MIT researcher who explored bias in face recognition services.

GARY MARCUS, the NYU psychologist who founded a start-up called Geometric Intelligence and sold it to Uber.

DEAN POMERLEAU, the Carnegie Mellon graduate student who used a neural network in building a driverless car in the late '80s and early '90s.

JÜRGEN SCHMIDHUBER, the researcher at the Dalle Molle Institute for Artificial Intelligence Research in Switzerland whose ideas helped drive the rise of deep learning.

TERRY SEJNOWSKI, the Johns Hopkins neuroscientist who was part of the neural network revival in the '80s.

AT THE SINGULARITY SUMMIT

PETER THIEL, the PayPal founder and early Facebook investor who met the founders of DeepMind at the Singularity Summit, a conference dedicated to futurism.

ELIEZER YUDKOWSKY, the futurist who introduced the DeepMind founders to Thiel.

IN THE PAST

MARVIN MINSKY, the AI pioneer who questioned Frank Rosenblatt's work and helped push it into the shadows.

FRANK ROSENBLATT, the Cornell psychology professor who built the Perceptron, a system that learned to recognize images, in the early 1960s.

DAVID RUMELHART, the University of California–San Diego psychologist and mathematician who helped revive Frank Rosenblatt's ideas alongside Geoff Hinton in the 1980s.

ALAN TURING, the founding father of the computer age who lived on the staircase at King's College Cambridge that was later home to Geoff Hinton.

NOTES

This book is based on interviews with more than four hundred people over the course of the eight years I've been reporting on artificial intelligence for *Wired* magazine and then the *New York Times*, as well as more than a hundred interviews conducted specifically for the book. Most people have been interviewed more than once, some several times or more. The book also draws on many company and personal documents and emails that reveal or corroborate particular events or details. Each scene and each significant detail (an acquisition price, for instance) was corroborated by at least two sources, often more. In the notes that follow, as a courtesy, I include references to my own stories published by former employers, including *Wired*. I include references to stories I and my collaborators have written at the *New York Times* only if the stories are explicitly mentioned in the narrative of this book—or if I just want to show appreciation for my collaborators. The book draws on all the same interviews and notes as my work at the *New York Times*.

PROLOGUE: THE MAN WHO DIDN'T SIT DOWN

2 **And its website offered nothing but a name:** Internet Archive, Web crawl from November 28, 2012, http://web.archive.org.

3 **it could identify common objects:** Alex Krizhevsky, Ilya Sutskever, Geoffrey Hinton, "ImageNet Classification with Deep Convolutional Neural Networks," *Advances in Neural Information Processing Systems* 25 (NIPS 2012), https://papers.nips.cc/paper/4824-imagenet-classification-with-deep-convolutional-neural-networks.pdf.

CHAPTER 1: GENESIS

15 **On July 7, 1958, several men:** "New Navy Device Learns by Doing," *New York Times,* July 8, 1958.

16 **this system would learn to recognize:** "Electronic 'Brain' Teaches Itself," *New York Times,* July 13, 1958.

16 **"The Navy revealed the embryo":** "New Navy Device Learns by Doing."

16 **A second article:** "Electronic 'Brain' Teaches Itself."

16 **Rosenblatt grew to resent:** Frank Rosenblatt, *Principles of Neurodynamics: Perceptrons and the Theory of Brain Mechanisms* (Washington, D.C: Spartan Books, 1962), pp. vii–viii.

17 **Frank Rosenblatt was born:** "Dr. Frank Rosenblatt Dies at 43; Taught Neurobiology at Cornell," *New York Times,* July 13, 1971.

17 **He attended Bronx Science:** "Profiles, AI, Marvin Minsky," *New Yorker,* December 14, 1981.

17 **eight Nobel laureates:** Andy Newman, "Lefkowitz is 8th Bronx Science H.S. Alumnus to Win Nobel Prize," *New York Times,* October 10, 2012, https://cityroom.blogs.nytimes.com/2012/10/10/another-nobel-for-bronx-science-this-one-in-chemistry/.

17 **six Pulitzer Prize winners, eight National Medal of Science winners:** Robert Wirsing, "Cohen Co-names 'Bronx Science Boulevard,'" *Bronx Times,* June 7, 2010, https://www.bxtimes.com/cohen-co-names-bronx-science-boulevard/.

17 **three recipients of the Turing Award:** The Bronx High School of Science website, "Hall of Fame," https://www.bxscience.edu/halloffame/; "Martin Hellman (Bronx Science Class of '62) Wins the A.M. Turing Award," The Bronx High School of Science website, https://www.bxscience.edu/m/news/show_news.jsp?REC_ID=403749&id=1.

17 **In 1953, the *New York Times* published a small story:** "Electronic Brain's One-Track Mind," *New York Times,* October 18, 1953.

17 **he joined the Cornell Aeronautical Laboratory:** "Dr. Frank Rosenblatt Dies at 43; Taught Neurobiology at Cornell."

17 **Rosenblatt saw the project:** Rosenblatt, *Principles of Neurodynamics: Perceptrons and the Theory of Brain Mechanisms,* pp. v–viii.

17 **If he could re-create the brain as a machine:** Ibid.

18 **Rosenblatt showed off the beginnings of this idea:** "New Navy Device Learns by Doing."

18 **"For the first time, a non-biological system":** "Rival," Talk of the Town, *New Yorker,* December 6, 1958.

18 **"My colleague disapproves":** Ibid.

19 **It lacked depth perception:** Ibid.

19 **"Love. Hope. Despair":** Ibid.

19 **Now it described the Perceptron:** Ibid.

19 **Though scientists claimed that only biological systems:** Ibid.

19 **"It is only a question":** Ibid.

19 **Rosenblatt completed the Mark I:** John Hay, Ben Lynch, and David Smith, *Mark I Perceptron Operators' Manual,* 1960, https://apps.dtic.mil/dtic/tr/fulltext/u2/236965.pdf.

21 **Frank Rosenblatt and Marvin Minsky had been contemporaries:** "Profiles, AI, Marvin Minsky."

21 **But he complained that neither could match:** Ibid.

21 **As an undergraduate at Harvard:** Stuart Russell and Peter Norvig, *Artificial Intelligence: A Modern Approach* (Upper Saddle River, NJ: Prentice Hall, 2010), p. 16.

21 **as a graduate student in the early '50s:** Marvin Minsky, *Theory of Neural-Analog Reinforcement Systems and Its Application to the Brain Model Problem* (Princeton, NJ: Princeton University, 1954).

21 **He was among the small group of scientists:** Russell and Norvig, *Artificial Intelligence: A Modern Approach,* p. 17.

21 **A Dartmouth professor named John McCarthy:** Claude Shannon and John McCarthy, *Automata Studies,* Annals of Mathematics Studies, April 1956 (Princeton, NJ: Princeton University Press).

21–22 **The agenda at the Dartmouth Summer Research Conference on Artificial Intelligence:** John McCarthy, Marvin Minsky, Nathaniel Rochester, and Claude Shannon, "A Proposal for the Dartmouth Summer Research Project on Artificial Intelligence," August 31, 1955, http://raysolomonoff.com/dartmouth/boxa/dart564props.pdf.

22 **they were sure it wouldn't take very long:** Herbert Simon and Allen Newell, "Heuristic Problem Solving: The Next Advance in Operations Research," *Operations Research* 6, no. 1 (January–February 1958), p. 7.

22 **the Perceptron was a controversial concept:** Rosenblatt, *Principles of Neurodynamics: Perceptrons and the Theory of Brain Mechanisms,* pp. v–viii.

22 **The reporters who wrote about his work in the late 1950s:** Ibid.

22–23 **"The perceptron program is not primarily concerned":** Ibid.

23 **In 1966, a few dozen researchers traveled to Puerto Rico:** Laveen Kanal, ed., *Pattern Recognition* (Washington, D.C.: Thompson Book Company, 1968), p. vii.

25 **published a book on neural networks:** Marvin Minsky and Seymour Papert, *Perceptrons* (Cambridge, MA: MIT Press, 1969).

26 **The movement reached the height of its ambition:** Cade Metz, "One Genius' Lonely Crusade to Teach a Computer Common Sense," *Wired*, March 24, 2016, https://www.wired.com/2016/03/doug-lenat-artificial-intelligence-common-sense-engine/.

26 **Rosenblatt shifted to a very different area of research:** "Dr. Frank Rosenblatt Dies at 43; Taught Neurobiology at Cornell."

CHAPTER 2: PROMISE

30 **George Boole, the nineteenth-century British mathematician and philosopher:** Desmond McHale, *The Life and Work of George Boole: A Prelude to the Digital Age* (Cork, Ireland: Cork University Press, 2014).

30 **James Hinton, the nineteenth-century surgeon:** Gerry Kennedy, *The Booles and the Hintons* (Cork, Ireland: Atrium Press, 2016).

30 **Charles Howard Hinton, the mathematician and fantasy writer:** Ibid.

30 **Sebastian Hinton invented the jungle gym:** U.S. Patent 1,471,465; U.S. Patent 1,488,244; U.S. Patent 1,488,245 1920; and U.S. Patent 1,488,246.

30 **the nuclear physicist Joan Hinton:** William Grimes, "Joan Hinton, Physicist Who Chose China over Atom Bomb, Is Dead at 88," *New York Times*, June 11, 2010, https://www.nytimes.com/2010/06/12/science/12hinton.html.

30 **the entomologist Howard Everest Hinton:** George Salt, "Howard Everest Hinton. 24 August 1912–2 August 1977," *Biographical Memoirs of Fellows of the Royal Society* (London: Royal Society Publishing, 1978), pp. 150–182, https://royalsocietypublishing.org/doi/10.1098/rsbm.1978.0006.

30 **Sir George Everest, the surveyor general of India:** Kennedy, *The Booles and the Hintons*.

31–32 **aimed to explain the basic biological process:** Peter M. Milner and Brenda Atkinson Milner, "Donald Olding Hebb. 22 July 1904–20 August 1985," *Biographical Memoirs of Fellows of the Royal Society* (London: Royal Society Publishing, 1996), 42: 192–204, https://royalsocietypublishing.org/doi/10.1098/rsbm.1996.0012.

32 **helped inspire the artificial neural networks:** Stuart Russell and Peter Norvig, *Artificial Intelligence: A Modern Approach* (Upper Saddle River, NJ: Prentice Hall, 2010), p. 16.

33 **Longuet-Higgins had been a theoretical chemist:** Chris Darwin, "Christopher Longuet-Higgins, Cognitive Scientist with a Flair for Chemistry," *Guardian*, June 9, 2004, https://www.theguardian.com/news/2004/jun/10/guardianobituaries.highereducation.

34 **the British government commissioned a study:** James Lighthill, "Artificial Intelligence: A General Survey," Artificial Intelligence: A Paper Symposium, Science Research Council, Great Britain, 1973.

34 **"Most workers in AI research":** Ibid.

35 **he published a call to arms:** Francis Crick, "Thinking About the Brain," *Scientific American,* September 1979.

39 **When anyone asked Feynman:** Lee Dye, "Nobel Physicist R. P. Feynman Dies," *Los Angeles Times,* February 16, 1988, https://www.latimes.com/archives/la-xpm-1988-02-16-mn-42968-story.html.

42 **It was published later that year:** David Rumelhart, Geoffrey Hinton, and Ronald Williams, "Learning Representations by Back-Propagating Errors," *Nature* 323 (1986), pp. 533–536.

44 **Created in 1958 in response to the Sputnik satellite:** "About DARPA," Defense Advanced Research Projects Agency website, https://www.darpa.mil/about-us/about-darpa.

44 **when Reagan administration officials secretly sold arms:** Lee Hamilton and Daniel Inouye, "Report of the Congressional Committees Investigating the Iran-Contra Affair" (Washington, D.C.: Government Printing Office, 1987).

CHAPTER 3: REJECTION

46 **Yann LeCun sat at a desktop computer:** "Convolutional Neural Network Video from 1993 [*sic*]," YouTube, https://www.youtube.com/watch?v=FwFduRA_L6Q.

48 **In October 1975, at the Abbaye de Royaumont:** Jean Piaget, Noam Chomsky, and Massimo Piattelli-Palmarini, *Language and Learning: The Debate Between Jean Piaget and Noam Chomsky* (Cambridge: Harvard University Press, 1980).

49 **Sejnowski was making waves with something he called "NETtalk":** "Learning, Then Talking," *New York Times,* August 16, 1988.

51 **His breakthrough was a variation:** Yann LeCun, Bernhard Boser, John Denker et al., "Backpropagation Applied to Handwritten Zip Code Recognition," *Neural Computation* (Winter 1989), http://yann.lecun.com/exdb/publis/pdf/lecun-89e.pdf.

52 **ANNA was the acronym for:** Eduard Säckinger, Bernhard Boser, Jane Bromley et al., "Application of the ANNA Neural Network Chip to High-Speed Character Recognition," *IEEE Transaction on Neural Networks* (March 1992).

53 **In 1995, two Bell Labs researchers:** Daniela Hernandez, "Facebook's Quest to Build an Artificial Brain Depends on This Guy," *Wired,* August 14, 2014, https://www.wired.com/2014/08/deep-learning-yann-lecun/.

54 **With a few wistful words:** http://yann.lecun.com/ex/group/index.html, retrieved March 9, 2020.

58 **One year, Clément Farabet:** Clément Farabet, Camille Couprie, Laurent Najman, and Yann LeCun, "Scene Parsing with Multiscale Feature Learn-

ing, Purity Trees, and Optimal Covers," 29th International Conference on Machine Learning (ICML 2012), June 2012, https://arxiv.org/abs/1202.2160.

59 **As a child, Schmidhuber had told his younger brother:** Ashlee Vance, "This Man Is the Godfather the AI Community Wants to Forget," *Bloomberg Businessweek,* May 15, 2018, https://www.bloomberg.com/news/features/2018-05-15/google-amazon-and-facebook-owe-j-rgen-schmidhuber-a-fortune.

59 **from the age of fifteen:** Jürgen Schmidhuber's Home Page, http://people.idsia.ch/~juergen/, retrieved March 9, 2020.

59 **his ambitions interlocked:** Vance, "This Man Is the Godfather the AI Community Wants to Forget."

62 **A researcher named Aapo Hyvärinen:** Aapo Hyvärinen, "Connections Between Score Matching, Contrastive Divergence, and Pseudolikelihood for Continuous-Valued Variables," revised submission to IEEE TNN, February 21, 2007, https://www.cs.helsinki.fi/u/ahyvarin/papers/cdsm3.pdf.

CHAPTER 4: BREAKTHROUGH

67 **one of Deng's students wrote a thesis:** Khaled Hassanein, Li Deng, and M. I. Elmasry, "A Neural Predictive Hidden Markov Model for Speaker Recognition," SCA Workshop on Automatic Speaker Recognition, Identification, and Verification, April 1994, https://www.isca-speech.org/archive_open/asriv94/sr94_115.html.

68 **Hinton sent Deng another email:** Abdel-rahman Mohamed, George E. Dahl, and Geoffrey Hinton, "Deep Belief Networks for Phone Recognition," NIPS workshop on deep learning for speech recognition and related applications, 2009, https://www.cs.toronto.edu/~gdahl/papers/dbnPhoneRec.pdf.

72 **In 2005, three engineers:** "GPUs for Machine Learning Algorithms," Eighth International Conference on Document Analysis and Recognition (ICDAR 2005).

72 **a team at Stanford University:** Rajat Raina, Anand Madhavan, and Andrew Y. Ng, "Large-Scale Deep Unsupervised Learning Using Graphics Processors," Computer Science Department, Stanford University, 2009, http://robotics.stanford.edu/~ang/papers/icml09-LargeScaleUnsupervisedDeepLearningGPU.pdf.

CHAPTER 5: TESTAMENT

80–81 **the Google self-driving car:** John Markoff, "Google Cars Drive Themselves, in Traffic," *New York Times,* October 9, 2010, https://www.nytimes.com/2010/10/10/science/10google.html.

81 **He soon married another roboticist:** Evan Ackerman and Erico Guizz, "Robots Bring Couple Together, Engagement Ensues," *IEEE Spectrum*, March 31, 2014, https://spectrum.ieee.org/automaton/robotics/humanoids/engaging-with-robots.

82 **a 2004 book titled *On Intelligence*:** Jeff Hawkins with Sandra Blakeslee, *On Intelligence: How a New Understanding of the Brain Will Lead to the Creation of Truly Intelligent Machines* (New York: Times Books, 2004).

84 **Jeff Dean walked into the same microkitchen:** Gideon Lewis-Kraus, "The Great AI Awakening," *New York Times Magazine*, December 14, 2006, https://www.nytimes.com/2016/12/14/magazine/the-great-ai-awakening.html.

85 **he built a software tool:** Ibid.

85 **he was among the top DEC researchers:** Cade Metz, "If Xerox PARC Invented the PC, Google Invented the Internet," *Wired*, August 8, 2012, https://www.wired.com/2012/08/google-as-xerox-parc/.

88 **They built a system:** John Markoff, "How Many Computers to Identify a Cat? 16,000," *New York Times*, June 25, 2012, https://www.nytimes.com/2012/06/26/technology/in-a-big-network-of-computers-evidence-of-machine-learning.html.

88 **Drawing on the power:** Ibid.

88 **Ng, Dean, and Corrado published:** Quoc V. Le, Marc'Aurelio Ranzato, Rajat Monga et al., "Building High-level Features Using Large Scale Unsupervised Learning," 2012, https://arxiv.org/abs/1112.6209.

88 **The project also appeared:** Markoff, "How Many Computers to Identify a Cat? 16,000."

89 **merely agreed to spend the summer:** Lewis-Kraus, "The Great AI Awakening."

89 **He felt like an oddity:** Ibid.

89 **Wallis asks a government official:** *The Dam Busters*, directed by Michael Anderson, Associated British Pathé (UK), 1955.

90 **these used thousands of central processing units:** Le, Ranzato, Monga et al., "Building High-Level Features Using Large Scale Unsupervised Learning."

92 **ImageNet was an annual contest:** Olga Russakovsky, Jia Deng, Hao Su et al., "ImageNet Large Scale Visual Recognition Challenge," 2014, https://arxiv.org/abs/1409.0575.

95–96 **It was nearly twice as accurate:** Alex Krizhevsky, Ilya Sutskever, and Geoffrey Hinton. "ImageNet Classification with Deep Convolutional Neural Networks," *Advances in Neural Information Processing Systems* 25 (NIPS 2012), https://papers.nips.cc/paper/4824-imagenet-classification-with-deep-convolutional-neural-networks.pdf.

97 **Alfred Wegener first proposed:** Richard Conniff, "When Continental Drift Was Considered Pseudoscience," *Smithsonian*, June 2012, https://www.smithsonianmag.com/science-nature/when-continental-drift-was-considered-pseudoscience-90353214/.

97 **he developed a degenerative brain condition:** Benedict Carey, "David Rumelhart Dies at 68; Created Computer Simulations of Perception," *New York Times,* March 11, 2011.

CHAPTER 6: AMBITION

100–101 **He would soon set a world record:** John Markoff, "Parachutist's Record Fall: Over 25 Miles in 15 Minutes," *New York Times,* October 24, 2014.

101 **Two of them, Demis Hassabis and David Silver:** Cade Metz, "What the AI Behind AlphaGo Can Teach Us About Being Human," *Wired,* May 19, 2016, https://www.wired.com/2016/05/google-alpha-go-ai/.

102 **"Despite its rarefied image":** Archived "Diaries" from Elixir, https://archive.kontek.net/republic.strategyplanet.gamespy.com/d1.shtml.

102 **Hassabis won the world team championship in Diplomacy:** Steve Boxer, "Child Prodigy Stands by Originality," *Guardian,* September 9, 2004, https://www.theguardian.com/technology/2004/sep/09/games.onlinesupplement.

103 **In his gap year:** David Rowan, "DeepMind: Inside Google's Super-Brain," *Wired UK,* June 22, 2015, https://www.wired.co.uk/article/deepmind.

103 **he kept a running online diary:** Archived "Diaries" from Elixir, https://archive.kontek.net/republic.strategyplanet.gamespy.com/d1.shtml.

104 **"Ian's no mean player":** Ibid.

104 **David Silver also returned to academia:** Metz, "What the AI Behind AlphaGo Can Teach Us About Being Human."

105 **With one paper, he studied people who developed amnesia:** Demis Hassabis, Dharshan Kumaran, Seralynne D. Vann, and Eleanor A. Maguire, "Patients with Hippocampal Amnesia Cannot Imagine New Experiences," *Proceedings of the National Academy of Sciences* 104, no. 5 (2007): pp. 1726–1731.

105 **In 2007, *Science*:** "Breakthrough of the Year," *Science,* December 21, 2007.

106 **Superintelligence, he had said in his thesis:** Shane Legg, "Machine Super Intelligence," PhD dissertation, University of Lugano, June 2008, http://www.vetta.org/documents/Machine_Super_Intelligence.pdf.

106 **"If one accepts that the impact of truly intelligent machines":** Ibid.

108 **In the summer of 2010, Hassabis and Legg arranged to address the Singularity Summit:** Hal Hodson, "DeepMind and Google: The Battle to Control Artificial Intelligence," *1843 Magazine,* April/May 2019, https://www.1843magazine.com/features/deepmind-and-google-the-battle-to-control-artificial-intelligence.

109 **He called this "the biological approach":** "A Systems Neuroscience Approach to Building AGI—Demis Hassabis, Singularity Summit 2010," YouTube, https://www.youtube.com/watch?v=Qgd3OK5DZWI.

109 **"We should be focusing on the algorithmic level":** Ibid.

109 **He told his audience that artificial intelligence researchers needed definitive ways:** "Measuring Machine Intelligence—Shane Legg, Singularity Summit," YouTube, https://www.youtube.com/watch?v=0ghzG14dT-w.

109 **"I want to know where we're going":** Ibid.

109 **Hassabis started talking chess:** Metz, "What the AI Behind AlphaGo Can Teach Us About Being Human."

110 **he invested £1.4 million of the initial £2 million:** Hodson, "DeepMind and Google: The Battle to Control Artificial Intelligence."

111 **when computer scientists built the first automated chess players:** Stuart Russell and Peter Norvig, *Artificial Intelligence: A Modern Approach* (Upper Saddle River, NJ: Prentice Hall, 2010), p. 14.

111 **In 1990, researchers marked a turning point:** Ibid., p. 186.

111 **Seven years later, IBM's Deep Blue supercomputer:** Ibid., p. ix.

111 **And in 2011, another IBM machine, Watson:** John Markoff, "Computer Wins on 'Jeopardy!': Trivial, It's Not," *New York Times,* February 16, 2011.

111 **This neural network could master the game:** Volodymyr Mnih, Koray Kavukcuoglu, David Silver et al., "Playing Atari with Deep Reinforcement Learning," *Nature* 518 (2015), pp. 529–533.

116 **it was acquiring DeepMind:** Samuel Gibbs, "Google Buys UK Artificial Intelligence Startup Deepmind for £400m," *Guardian,* January 27, 2017, https://www.theguardian.com/technology/2014/jan/27/google-acquires-uk-artificial-intelligence-startup-deepmind.

CHAPTER 7: RIVALRY

120 **That night, Facebook held a private party:** "Facebook Buys into Machine Learning," Neil Lawrence blog and video, https://inverseprobability.com/2013/12/09/facebook-buys-into-machine-learning.

120 **"It's a marriage made in heaven":** Ibid.

128 **It even shared the designs:** Cade Metz, "Facebook 'Open Sources' Custom Server and Data Center Designs," *Register,* April 7, 2011, https://www.theregister.co.uk/2011/04/07/facebook_data_center_unveiled/.

129 **Even old hands, like Jeff Dean:** Cade Metz, "Google Just Open Sourced TensorFlow, Its Artificial Intelligence Engine," *Wired,* November 9, 2015, https://www.wired.com/2015/11/google-open-sources-its-artificial-intelligence-engine/.

130 **Rick Rashid, Microsoft's head of research:** John Markoff, "Scientists See Promise in Deep-Learning Programs Image," *New York Times,* November 23, 2012, https://www.nytimes.com/2012/11/24/science/scientists-see-advances-in-deep-learning-a-part-of-artificial-intelligence.html.

132 **its staff costs totaled $260 million:** DeepMind Technologies Limited Report and Financial Statements Year Ended, December 31, 2017.

132 **As Microsoft vice president Peter Lee told *Bloomberg Businessweek*:** Ashlee Vance, "The Race to Buy the Human Brains Behind Deep Learning Machines," *Bloomberg Businessweek,* January 27, 2014, https://www.bloomberg.com/news/articles/2014-01-27/the-race-to-buy-the-human-brains-behind-deep-learning-machines.

132 **Andrew Ng would be running labs in both Silicon Valley and Beijing:** Daniela Hernandez, "Man Behind the 'Google Brain' Joins Chinese Search Giant Baidu," *Wired,* May 16, 2014, https://www.wired.com/2014/05/andrew-ng-baidu/.

CHAPTER 8: HYPE

133 **Felix Baumgartner soon set a world skydiving record:** John Tierney, "24 Miles, 4 Minutes and 834 M.P.H., All in One Jump," *New York Times,* October 14, 2012, https://www.nytimes.com/2012/10/15/us/felix-baumgartner-skydiving.html.

136 **he broke Baumgartner's skydiving record:** John Markoff, "Parachutist's Record Fall: Over 25 Miles in 15 Minutes," *New York Times,* October 24, 2014, https://www.nytimes.com/2014/10/25/science/alan-eustace-jumps-from-stratosphere-breaking-felix-baumgartners-world-record.html.

137 **The Chauffeur engineers called him "the AI whisperer":** Andrew J. Hawkins, "Inside Waymo's Strategy to Grow the Best Brains for Self-Driving Cars," *The Verge,* May 9, 2018, https://www.theverge.com/2018/5/9/17307156/google-waymo-driverless-cars-deep-learning-neural-net-interview.

138 **the company's $56 billion in annual revenue:** Google Annual Report, 2013, https://www.sec.gov/Archives/edgar/data/1288776/000128877614000020/goog2013123110-k.htm.

139 **they unveiled a system called RankBrain:** Jack Clark, "Google Turning Its Lucrative Web Search Over to AI Machines," *Bloomberg News,* October 26, 2015, https://www.bloomberg.com/news/articles/2015-10-26/google-turning-its-lucrative-web-search-over-to-ai-machines.

139 **It helped drive about 15 percent:** Ibid.

139 **Singhal left the company:** Cade Metz, "AI Is Transforming Google Search. The Rest of the Web Is Next," *Wired,* February 4, 2016, https://www.wired.com/2016/02/ai-is-changing-the-technology-behind-google-searches/; Mike Isaac and Daisuke Wakabayashi, "Amit Singhal, Uber Executive Linked to Old Harassment Claim, Resigns," *New York Times,* February 27, 2017, https://www.nytimes.com/2017/02/27/technology/uber-sexual-harassment-amit-singhal-resign.html.

139 **he was replaced as the head of Google Search:** Metz, "AI Is Transforming Google Search. The Rest of the Web Is Next."

139 **In London, Demis Hassabis soon revealed:** Jack Clarke, "Google Cuts Its Giant Electricity Bill with DeepMind-Powered AI," *Bloomberg News,* July 19, 2016, https://www.bloomberg.com/news/articles/2016-07-19/google-cuts-its-giant-electricity-bill-with-deepmind-powered-ai.

139 **This system decided when to turn on:** Ibid.

139 **The Google data centers were so large:** Ibid.

141 **automated technologies would soon cut a giant swath through the job market:** Carl Benedikt Frey and Michael A. Osborne, "The Future of Employment: How Susceptible Are Jobs to Computerisation?" Working paper, Oxford Martin School, September 2013, https://www.oxfordmartin.ox.ac.uk/downloads/academic/The_Future_of_Employment.pdf.

142 **"You have been Schmidhubered":** Ashlee Vance, "This Man Is the Godfather the AI Community Wants to Forget," *Bloomberg Businessweek,* May 15, 2018, https://www.bloomberg.com/news/features/2018-05-15/google-amazon-and-facebook-owe-j-rgen-schmidhuber-a-fortune.

144 **Google Brain had already explored a technology called "word embeddings":** Tomas Mikolov, Ilya Sutskever, Kai Chen et al., "Distributed Representations of Words and Phrases and their Compositionality," 2013, https://arxiv.org/abs/1301.3781.

144 **Sutskever's translation system was an extension of this idea:** Ilya Sutskever, Oriol Vinyals, and Quoc V. Le, "Sequence to Sequence Learning with Neural Networks," 2014, https://arxiv.org/abs/1409.3215.

145 **Sutskever presented a paper:** "NIPS Oral Session 4—Ilya Sutskever," YouTube, https://www.youtube.com/watch?v=-uyXE7dY5H0.

146 **Google already operated more than fifteen data centers:** Cade Metz, "Building an AI Chip Saved Google from Building a Dozen New Data Centers," *Wired,* April 5, 2017, https://www.wired.com/2017/04/building-ai-chip-saved-google-building-dozen-new-data-centers/.

147 **Google had a long history of building its own data center hardware:** Cade Metz, "Revealed: The Secret Gear Connecting Google's Online Empire," *Wired,* June 17, 2015, https://www.wired.com/2015/06/google-reveals-secret-gear-connects-online-empire/.

147 **as Facebook, Amazon, and others followed suit:** Robert McMillan and Cade Metz, "How Amazon Followed Google into the World of Secret Servers," *Wired,* November 30, 2012, https://www.wired.com/2012/11/amazon-google-secret-servers/.

148 **The trick was that its calculations were *less precise*:** Gideon Lewis-Kraus, "The Great AI Awakening," *New York Times Magazine,* December 14, 2006, https://www.nytimes.com/2016/12/14/magazine/the-great-ai-awakening.html.

148 **Their dataset was somewhere between a hundred and a thousand times larger:** Ibid.

148 **Dean tapped three engineers:** Ibid.

149 **For English and French, its BLEU score:** Ibid.

149 **built a neural network that topped the existing system by seven points:** Ibid.

149 **It took ten seconds to translate a ten-word sentence:** Ibid.

149 **Hughes thought the company would need three years:** Ibid.

149 **Dean, however, thought otherwise:** Ibid.

149 **"We can do it by the end of the year, if we put our minds to it":** Ibid.

149 **Hughes was skeptical:** Ibid.

149 **"I'm not going to be the one":** Ibid.

149 **The Chinese Internet giant had published a paper describing similar research:** Ibid.

150 **A sentence that needed ten seconds:** Ibid.

150 **They released the first incarnation of the service just after Labor Day:** Ibid.

150 **he and Jeff Dean worked on a project they called "Distillation":** Geoffrey Hinton, Oriol Vinyals, and Jeff Dean, "Distilling the Knowledge in a Neural Network," 2015, https://arxiv.org/abs/1503.02531.

CHAPTER 9: ANTI-HYPE

152 **On November 14, 2014, Elon Musk posted a message:** James Cook, "Elon Musk: You Have No Idea How Close We Are to Killer Robots," *Business Insider UK,* November 17, 2014, https://www.businessinsider.com/elon-musk-killer-robots-will-be-here-within-five-years-2014-11.

153 **Musk said his big fear:** Ashlee Vance, *Elon Musk: Tesla, SpaceX, and the Quest for a Fantastic Future* (New York: Ecco, 2017).

153 **The trouble was not that Page:** Ibid.

153 **The trouble was that Page operated:** Ibid.

153 **"He could produce something evil by accident":** Ibid.

153 **he invoked *The Terminator*:** "Closing Bell," CNBC, transcript, https://www.cnbc.com/2014/06/18/first-on-cnbc-cnbc-transcript-spacex-ceo-elon-musk-speaks-with-cnbcs-closing-bell.html.

153 **"potentially more dangerous than nukes":** Elon Musk tweet, August 2, 2014, https://twitter.com/elonmusk/status/495759307346952192?s=19.

153 **The same tweet urged his followers to read *Superintelligence*:** Ibid.

153 **Bostrom believed that superintelligence:** Nick Bostrom, *Superintelligence: Paths, Dangers, Strategies* (Oxford, UK: Oxford University Press, 2014).

153 **"This is quite possibly the most important":** Ibid.

154 **warning author Walter Isaacson about the dangers of artificial intelligence:** Lessley Anderson, "Elon Musk: A Machine Tasked with Getting Rid

of Spam Could End Humanity," *Vanity Fair,* October 8, 2014, https://www.vanityfair.com/news/tech/2014/10/elon-musk-artificial-intelligence-fear.

154 **If researchers designed a system to fight email spam:** Ibid.

154 **When Isaacson asked if he would use his SpaceX rockets:** Ibid.

154 **"If there's some apocalypse scenario":** Ibid.

154 **Musk posted his message to Edge.org:** Cook, "Elon Musk: You Have No Idea How Close We Are to Killer Robots."

154 **the billionaire Jeffrey Epstein:** William K. Rashbaum, Benjamin Weiser, and Michael Gold, "Jeffrey Epstein Dead in Suicide at Jail, Spurring Inquiries," *New York Times,* August 10, 2019, https://www.nytimes.com/2019/08/10/nyregion/jeffrey-epstein-suicide.html.

154 **He pointed to DeepMind:** Cook, "Elon Musk: You Have No Idea How Close We Are to Killer Robots."

154 **He said danger was five to ten years away:** Ibid.

157 **Shane Legg described this attitude in his thesis:** Shane Legg, "Machine Super Intelligence," 2008, http://www.vetta.org/documents/Machine_Super_Intelligence.pdf.

157 **"If there is ever to be something approaching absolute power":** Ibid.

158 **when it invited this growing community to a private summit:** Max Tegmark, *Life 3.0: Being Human in the Age of Artificial Intelligence* (New York: Random House, 2017).

158 **it aimed to create a meeting of the minds along the lines of the Asilomar conference:** Ibid.

158 **Musk took the stage to discuss the threat of an intelligence explosion:** Robert McMillan, "AI Has Arrived, and That Really Worries the World's Brightest Minds," *Wired,* January 16, 2015, https://www.wired.com/2015/01/ai-arrived-really-worries-worlds-brightest-minds/.

158 **That, he said, was the big risk:** Ibid.

159 **Musk pledged $10 million:** Tegmark, *Life 3.0: Being Human in the Age of Artificial Intelligence.*

159 **But as he prepared to announce this new gift:** Ibid.

159 **Someone reminded him there were no reporters:** Ibid.

159 **So he made the announcement without mentioning the dollar figure:** Ibid.

159 **he revealed the $10 million grant in a tweet:** Elon Musk tweet, January 15, 2015, https://twitter.com/elonmusk/status/555743387056226304.

159 **Tegmark distributed an open letter:** "An Open Letter, Research Priorities for Robust and Beneficial Artificial Intelligence," Future of Life Institute, https://futureoflife.org/ai-open-letter/.

159 **"We believe that research on how to make AI systems robust":** Ibid.

160 **One person who attended the conference but did not sign was Kent Walker:** Ibid.

160 **one of the top researchers inside Google Brain:** Ibid.

160 **Max Tegmark later wrote a book about the potential impact of superintelligence:** Tegmark, *Life 3.0: Being Human in the Age of Artificial Intelligence.*
160 **In the opening pages, he described:** Ibid.
160 **Page mounted a defense of what Tegmark described as "digital utopianism":** Ibid.
160 **Page worried that paranoia:** Ibid.
160 **Musk pushed back, asking how Page could be sure this superintelligence:** Ibid.
162 **Brockman vowed to build the new lab they all seemed to want:** Cade Metz, "Inside OpenAI, Elon Musk's Wild Plan to Set Artificial Intelligence Free," *Wired,* April 27, 2016, https://www.wired.com/2016/04/openai-elon-musk-sam-altman-plan-to-set-artificial-intelligence-free/.
164 **nearly $2 million for the first year:** OpenAI, form 990, 2016.
164 **Musk and Altman painted OpenAI as a counterweight:** Steven Levy, "How Elon Musk and Y Combinator Plan to Stop Computers from Taking Over," "Backchannel," *Wired,* December 11, 2015, https://www.wired.com/2015/12/how-elon-musk-and-y-combinator-plan-to-stop-computers-from-taking-over/.
165 **backed by over a billion dollars in funding:** Ibid.
165 **AI would be available to everyone:** Ibid.
165 **if they open-sourced all their research, the bad actors:** Ibid.
165 **But they argued that the threat of malicious AI:** Ibid.
165 **"We think it's far more likely that many, many AIs":** Ibid.
165 **said those "borderline-crazy":** Metz, "Inside OpenAI, Elon Musk's Wild Plan to Set Artificial Intelligence Free."

CHAPTER 10: EXPLOSION

167 **On October 31, 2015, Facebook chief technology officer Mike Schroepfer:** Cade Metz, "Facebook Aims Its AI at the Game No Computer Can Crack," *Wired,* November 3, 2015, https://www.wired.com/2015/11/facebook-is-aiming-its-ai-at-go-the-game-no-computer-can-crack/.
167 **about a French computer scientist:** Alan Levinovitz, "The Mystery of Go, the Ancient Game That Computers Still Can't Win," *Wired,* May 12, 2014, https://www.wired.com/2014/05/the-world-of-computer-go/.
168 **Facebook researchers were confident they could crack the game:** Metz, "Facebook Aims Its AI at the Game No Computer Can Crack."
168 **"The best players end up looking at visual patterns":** Ibid.
169 **It was analyzing photos and generating captions:** Cade Metz, "Facebook's AI Is Now Automatically Writing Photo Captions," *Wired,* April 5, 2016, https://www.wired.com/2016/04/facebook-using-ai-write-photo-captions-blind-users/.

169 **It was driving Facebook M:** Cade Metz, "Facebook's Human-Powered Assistant May Just Supercharge AI," *Wired,* August 26, 2015, https://www.wired.com/2015/08/how-facebook-m-works/.

169 **built a system that could read passages from *The Lord of the Rings*:** Metz, "Facebook Aims Its AI at the Game No Computer Can Crack."

169 **Demis Hassabis appeared in an online video:** "Interview with Demis Hassabis," YouTube, https://www.youtube.com/watch?v=EhAjLnT9aL4.

169 **"I can't talk about it yet":** Ibid.

170 **Hassabis and DeepMind revealed that their AI system, AlphaGo:** Cade Metz, "In a Huge Breakthrough, Google's AI Beats a Top Player at the Game of Go," *Wired,* January 27, 2016, https://www.wired.com/2016/01/in-a-huge-breakthrough-googles-ai-beats-a-top-player-at-the-game-of-go/.

170 **Demis Hassabis and several other DeepMind researchers:** Cade Metz, "What the AI Behind AlphaGo Can Teach Us About Being Human," *Wired,* May 19, 2016, https://www.wired.com/2016/05/google-alpha-go-ai/.

171 **The four researchers published a paper on their early work around the middle of 2014:** Chris J. Maddison, Aja Huang, Ilya Sutskever, and David Silver, "Move Evaluation in Go Using Deep Convolutional Neural Networks," 2014, https://arxiv.org/abs/1412.6564.

172 **Over 200 million people would watch AlphaGo versus Lee Sedol:** https://deepmind.com/research/case-studies/alphago-the-story-so-far.

172 **put Lee in an entirely different echelon of the game:** Cade Metz, "Google's AI Is About to Battle a Go Champion—But This Is No Game," *Wired,* March 8, 2016, http://wired.com/2016/03/googles-ai-taking-one-worlds-top-go-players/.

172 **He and his team originally taught the machine to play Go by feeding 30 million moves:** Metz, "What the AI Behind AlphaGo Can Teach Us About Being Human."

173 **Google's $75 billion Internet business:** Google Annual Report, 2015, https://www.sec.gov/Archives/edgar/data/1288776/000165204416000012/goog10-k2015.htm.

173 **Sergey Brin flew into Seoul:** Metz, "What the AI Behind AlphaGo Can Teach Us About Being Human."

173 **This room was filled with PCs and laptops:** Cade Metz, "How Google's AI Viewed the Move No Human Could Understand," *Wired,* March 14, 2016, https://www.wired.com/2016/03/googles-ai-viewed-move-no-human-understand.

173 **a team of Google engineers had run their own ultra-high-speed fiber-optic cable:** Cade Metz, "Go Grandmaster Lee Sedol Grabs Consolation Win Against Google's AI," *Wired,* March 13, 2016, https://www.wired.com/2016/03/go-grandmaster-lee-sedol-grabs-consolation-win-googles-ai/.

173 **"I can't tell you how tense it is":** Ibid.

174 **told the world he was in shock:** Cade Metz, "Go Grandmaster Says He's 'in Shock' but Can Still Beat Google's AI," *Wired,* March 9, 2016, https://www.wired.com/2016/03/go-grandmaster-says-can-still-beat-googles-ai/.

174 **"I don't really know if it's a good move or a bad move":** Ibid.

174 **"I thought it was a mistake":** Ibid.

175 **"It's not a human move":** Ibid.

175 **"It discovered this for itself":** Ibid.

176 **said that he, too, felt a sadness:** Cade Metz, "The Sadness and Beauty of Watching Google's AI Play Go," *Wired,* March 11, 2016, https://www.wired.com/2016/03/sadness-beauty-watching-googles-ai-play-go/.

176 **"There was an inflection point":** Ibid.

176 **Lee Sedol lost the third game:** Metz, "What the AI Behind AlphaGo Can Teach Us About Being Human."

176 **"I don't know what to say today":** Cade Metz, "In Two Moves, AlphaGo and Lee Sedol Redefined the Future," *Wired,* March 16, 2016, https://www.wired.com/2016/03/two-moves-alphago-lee-sedol-redefined-future/.

176 **Hassabis found himself hoping the Korean:** Metz, "What the AI Behind AlphaGo Can Teach Us About Being Human."

177 **"All the thinking that AlphaGo had done up to that point was sort of rendered useless":** Ibid.

177 **"I have improved already":** Ibid.

CHAPTER 11: EXPANSION

179 **nearly 70 million people are diabetic:** "Diabetes Epidemic: 98 Million People in India May Have Type 2 Diabetes by 2030," *India Today,* November 22, 2018, https://www.indiatoday.in/education-today/latest-studies/story/98-million-indians-diabetes-2030-prevention-1394158-2018-11-22.

180 **every 1 million people:** International Council of Ophthalmology, http://www.icoph.org/ophthalmologists-worldwide.html.

181 **Offering a $40,000 prize:** Merck Molecular Activity Challenge, https://www.kaggle.com/c/MerckActivity.

183 **Peng and her team acquired about one hundred and thirty thousand digital eye scans:** Varun Gulshan, Lily Peng, and Marc Coram, "Development and Validation of a Deep Learning Algorithm for Detection of Diabetic Retinopathy in Retinal Fundus Photographs," *JAMA* 316, no. 22 (January 2016), pp. 2402–2410, https://jamanetwork.com/journals/jama/fullarticle/2588763.

184 **Peng and her team acknowledged:** Cade Metz, "Google's AI Reads Retinas to Prevent Blindness in Diabetics," *Wired,* November 29, 2016, https://www.wired.com/2016/11/googles-ai-reads-retinas-prevent-blindness-diabetics/.

184 **"Don't believe anyone who says that it is":** Siddhartha Mukherjee, "AI Versus M.D.," *New Yorker*, March 27, 2017, https://www.newyorker.com/magazine/2017/04/03/ai-versus-md.

185 **"I think that if you work as a radiologist":** Ibid.

185 **He argued that neural networks would eclipse:** Ibid.

185 **would eventually provide a hitherto impossible level of healthcare:** Ibid.

185 **these algorithms would read X-rays, CAT scans, and MRIs:** Ibid.

185 **they would also make pathological diagnoses:** Ibid.

185 **"There's much more to learn here":** Ibid.

186 **Larry Page and Sergey Brin spun off several Google projects:** Conor Dougherty, "Google to Reorganize as Alphabet to Keep Its Lead as an Innovator," *New York Times,* August 10, 2015, https://www.nytimes.com/2015/08/11/technology/google-alphabet-restructuring.html.

187 **they found little common ground:** David Rowan, "DeepMind: Inside Google's Super-Brain," *Wired UK,* June 22, 2015, https://www.wired.co.uk/article/deepmind.

187 **Hassabis would propose complex:** Ibid.

187 **"We have to engage with the real world today":** Ibid.

187 **Suleyman unveiled what he called DeepMind Health:** Jordan Novet, "Google's DeepMind AI Group Unveils Health Care Ambitions," *Venturebeat,* February 24, 2016, https://venturebeat.com/2016/02/24/googles-deepmind-ai-group-unveils-heath-care-ambitions/.

188 **revealed the agreement between DeepMind:** Hal Hodson, "Revealed: Google AI has access to huge haul of NHS patient data," *New Scientist,* April 29, 2016, https://www.newscientist.com/article/2086454-revealed-google-ai-has-access-to-huge-haul-of-nhs-patient-data/.

188 **The deal gave DeepMind access to healthcare records for 1.6 million patients:** Ibid.

188 **a British regulator ruled:** Timothy Revell, "Google DeepMind's NHS Data Deal 'Failed to Comply' with Law," *New Scientist,* July 3, 2017, https://www.newscientist.com/article/2139395-google-deepminds-nhs-data-deal-failed-to-comply-with-law/.

CHAPTER 12: DREAMLAND

190 **spending no less than $7.6 billion to acquire Nokia:** Nick Wingfield, "Microsoft to Buy Nokia Units and Acquire Executive," *New York Times,* September 3, 2013, https://www.nytimes.com/2013/09/04/technology/microsoft-acquires-nokia-units-and-leader.html.

191 **describing the sweeping hardware and software system:** Jeffrey Dean, Greg S. Corrado, Rajat Monga et al., "Large Scale Distributed Deep Networks,"

Advances in Neural Information Processing Systems 25 (NIPS 2012), https://papers.nips.cc/paper/4687-large-scale-distributed-deep-networks.pdf.

192 **University of Washington professor Pedro Domingos called them "tribes":** Pedro Domingos, *The Master Algorithm: How the Quest for the Ultimate Learning Machine Will Remake Our World* (New York: Basic Books, 2015).

193 **an article in *Vanity Fair*:** Kurt Eichenwald, "Microsoft's Lost Decade," *Vanity Fair,* July 24, 2012, https://www.vanityfair.com/news/business/2012/08/microsoft-lost-mojo-steve-ballmer.

195 **Brought up by his grandfather:** Jennifer Bails, "Bing It On," *Carnegie Mellon Today,* October 1, 2010, https://www.cmu.edu/cmtoday/issues/october-2010-issue/feature-stories/bing-it-on/index.html.

197 **After Twitter acquired Madbits:** Catherine Shu, "Twitter Acquires Image Search Startup Madbits," *TechCrunch,* July 29, 2014, https://gigaom.com/2014/07/29/twitter-acquires-deep-learning-startup-madbits/.

197 **Uber bought a start-up called Geometric Intelligence:** Mike Isaac, "Uber Bets on Artificial Intelligence with Acquisition and New Lab," *New York Times,* December 5, 2016, https://www.nytimes.com/2016/12/05/technology/uber-bets-on-artificial-intelligence-with-acquisition-and-new-lab.html.

200 **Microsoft announced his departure in September 2016:** Kara Swisher and Ina Fried, "Microsoft's Qi Lu Is Leaving the Company Due to Health Issues Rajesh Jha Will Assume Many of Lu's Responsibilities," *Recode,* September 29, 2016, https://www.vox.com/2016/9/29/13103352/microsoft-qi-lu-to-exit.

200 **he returned to China and joined Baidu as chief operating officer:** "Microsoft Veteran Will Help Run Chinese Search Giant Baidu," *Bloomberg News,* January 16, 2017, https://www.bloomberg.com/news/articles/2017-01-17/microsoft-executive-qi-lu-departs-to-join-china-s-baidu-as-coo.

CHAPTER 13: DECEIT

205 **still slightly drunk:** Cade Metz, "Google's Dueling Neural Networks Spar to Get Smarter, No Humans Required," *Wired,* April 11, 2017, https://www.wired.com/2017/04/googles-dueling-neural-networks-spar-get-smarter-no-humans-required/.

205 **"My friends are wrong!":** Ibid.

205 **"If it hadn't worked, I might have given up on the idea":** Ibid.

206 **Yann LeCun called GANs "the coolest idea in deep learning in the last twenty years":** Davide Castelvecchi, "Astronomers Explore Uses for AI-Generated Images," *Nature,* February 1, 2017, https://www.nature.com/news/astronomers-explore-uses-for-ai-generated-images-1.21398.

206 **Researchers at the University of Wyoming built a system:** Anh Nguyen, Jeff Clune, Yoshua Bengio et al., "Plug & Play Generative Networks: Condi-

tional Iterative Generation of Images in Latent Space," 2016, https://arxiv.org/abs/1612.00005.

206 **A team at Nvidia built:** Ming-Yu Liu, Thomas Breuel, and Jan Kautz, "Unsupervised Image-to-Image Translation Networks," 2016, https://arxiv.org/abs/1703.00848.

206 **A group at the University of California–Berkeley designed a system:** Jun-Yan Zhu, Taesung Park, Phillip Isola, and Alexei A. Efros, "Unpaired Image-to-Image Translation using Cycle-Consistent Adversarial Networks," 2016, https://arxiv.org/abs/1703.10593.

207 **As the number of international students studying:** Lily Jackson, "International Graduate-Student Enrollments and Applications Drop for 2nd Year in a Row," *Chronicle of Higher Education,* February 7, 2019, https://www.chronicle.com/article/International-Graduate-Student/245624.

207 **Microsoft ended up buying Maluuba:** "Microsoft Acquires Artificial-Intelligence Startup Maluuba," *Wall Street Journal,* January 13, 2007, https://www.wsj.com/articles/microsoft-acquires-artificial-intelligence-startup-maluuba-1484338762.

207 **Hinton helped open the Vector Institute:** Steve Lohr, "Canada Tries to Turn Its AI Ideas into Dollars," *New York Times,* April 9, 2017, https://www.nytimes.com/2017/04/09/technology/canada-artificial-intelligence.html.

207 **It was backed by $130 million in funding:** Ibid.

207 **Prime Minister Justin Trudeau promised $93 million:** Ibid.

209 **saying it was a "pretty crazy idea":** Mike Isaac, "Facebook, in Cross Hairs After Election, Is Said to Question Its Influence," *New York Times,* November 12, 2016, https://www.nytimes.com/2016/11/14/technology/facebook-is-said-to-question-its-influence-in-election.html.

209 **with headlines like "FBI Agent Suspected":** Craig Silverman, "Here Are 50 of the Biggest Fake News Hits on Facebook From 2016," *Buzzfeed News,* December 30, 2016, https://www.buzzfeednews.com/article/craigsilverman/top-fake-news-of-2016.

209 **After Facebook revealed that a Russian company:** Scott Shane and Vindu Goel, "Fake Russian Facebook Accounts Bought $100,000 in Political Ads," *New York Times,* September 6, 2017, https://www.nytimes.com/2017/09/06/technology/facebook-russian-political-ads.html.

209 **A team from the University of Washington:** Supasorn Suwajanakorn, Steven Seitz, and Ira Kemelmacher-Shlizerman, "Synthesizing Obama: Learning Lip Sync from Audio," 2017, https://grail.cs.washington.edu/projects/AudioToObama/.

209 **engineers used similar techniques to turn Donald Trump:** Paul Mozur and Keith Bradsher, "China's A.I. Advances Help Its Tech Industry, and State Security," *New York Times,* December 3, 2017, https://www.nytimes.com/2017/12/03/business/china-artificial-intelligence.html.

210 **a team of researchers at a Nvidia lab in Finland unveiled:** Tero Karras, Timo Aila, Samuli Laine, and Jaakko Lehtinen, "Progressive Growing of GANs for Improved Quality, Stability, and Variation," 2017, https://arxiv.org/abs/1710.10196.

210 **Ian Goodfellow gave a speech at a small conference:** Jackie Snow, "AI Could Set Us Back 100 Years When It Comes to How We Consume News," *MIT Technology Review,* November 7, 2017, https://www.technologyreview.com/s/609358/ai-could-send-us-back-100-years-when-it-comes-to-how-we-consume-news/.

210 **He acknowledged that anyone:** Ibid.

210 **"We're speeding up things that are already possible":** Ibid.

211 **they would end the era:** Ibid.

211 **"It's been a little bit of a fluke":** Ibid.

211 **But that would be a hard transition:** Ibid.

211 **"Unfortunately, people these days are not very good at critical thinking":** Ibid.

211 **be a period of adjustment:** Ibid.

211 **"There's a lot of other areas where AI":** Ibid.

211 **someone calling themselves "Deepfakes":** Samantha Cole, "AI-Assisted Fake Porn Is Here and We're All Fucked," *Motherboard,* December 11, 2017, https://www.vice.com/en_us/article/gydydm/gal-gadot-fake-ai-porn.

211 **Services like Pornhub and Reddit:** Samantha Cole, "Twitter Is the Latest Platform to Ban AI-Generated Porn: Deepfakes Are in Violation of Twitter's Terms of Use," *Motherboard,* February 6, 2018, https://www.vice.com/en_us/article/ywqgab/twitter-bans-deepfakes; Arjun Kharpal, "Reddit, Pornhub Ban Videos that Use AI to Superimpose a Person's Face," CNBC, February 8, 2018, https://www.cnbc.com/2018/02/08/reddit-pornhub-ban-deepfake-porn-videos.html.

212 **he began to explore a separate technique:** Cade Metz, "How to Fool AI into Seeing Something That Isn't There," *Wired,* April 29, 2017, https://www.wired.com/2016/07/fool-ai-seeing-something-isnt/.

212 **Soon a team of researchers showed that by slapping a few Post-it notes:** Kevin Eykholt, Ivan Evtimov, Earlence Fernandes et al., "Robust Physical-World Attacks on Deep Learning Models," 2017, https://arxiv.org/abs/1707.08945.

212 **Goodfellow warned that the same phenomenon:** Metz, "How to Fool AI into Seeing Something That Isn't There."

212 **he was paid $800,000:** OpenAI, form 990, 2016.

CHAPTER 14: HUBRIS

214 **a two-hundred-thousand-square-foot conference center:** "Unveiling the Wuzhen Internet Intl Convention Center," *China Daily,* November 15, 2016, https://www.chinadaily.com.cn/business/2016-11/15/content_27381349.htm.

214 **Its roof spans more than 2.5 trillion tiles:** Ibid.

214 **Built to host the World Internet Conference:** Ibid.

215 **Demis Hassabis sat in a plush:** Cade Metz, "Google's AlphaGo Levels Up from Board Games to Power Grids," *Wired,* May 24, 2017, https://www.wired.com/2017/05/googles-alphago-levels-board-games-power-grids/.

215 **In 2010, Google had suddenly and dramatically:** Andrew Jacobs and Miguel Helft, "Google, Citing Attack, Threatens to Exit China," *New York Times,* January 12, 2010, https://www.nytimes.com/2010/01/13/world/asia/13beijing.html.

216 **There were more people on the Internet in China:** "Number of Internet Users in China from 2017 to 2023," *Statista,* https://www.statista.com/statistics/278417/number-of-internet-users-in-china/

216 **An estimated 60 million Chinese had watched the match against Lee Sedol:** "AlphaGo Computer Beats Human Champ in Hard-Fought Series," Associated Press, March 15, 2016, https://www.cbsnews.com/news/googles-alphago-computer-beats-human-champ-in-hard-fought-series/.

217 **With a private order sent to all Chinese media in Wuzhen:** Cade Metz, "Google Unleashes AlphaGo in China—But Good Luck Watching It There," *Wired,* May 23, 2017, https://www.wired.com/2017/05/google-unleashes-alphago-china-good-luck-watching/.

219 **Baidu opened its first outpost in Silicon Valley:** Daniela Hernandez, "'Chinese Google' Opens Artificial-Intelligence Lab in Silicon Valley," *Wired,* April 12, 2013, https://www.wired.com/2013/04/baidu-research-lab/.

219 **It was called the Institute of Deep Learning:** Ibid.

219 **Kai Yu told a reporter its aim was to simulate:** Ibid.

219 **"We are making progress day by day":** Ibid.

220 **Sitting in a chair next to his Chinese interviewer:** Cade Metz, "Google Is Already Late to China's AI Revolution," *Wired,* June 2, 2017, https://www.wired.com/2017/06/ai-revolution-bigger-google-facebook-microsoft/.

220 **"All of them would be better off":** Ibid.

221 **more than 90 percent of Google's revenues still came from online advertising:** Google Annual Report, 2016, https://www.sec.gov/Archives/edgar/data/1652044/000165204417000008/goog10-kq42016.htm.

221 **Amazon, whose cloud revenue would top $17.45 billion in 2017:** Amazon Annual Report, 2017, https://www.sec.gov/Archives/edgar/data/1018724/000101872419000004/amzn-20181231x10k.htm.

222 **born in Beijing before emigrating to the United States:** Octavio Blanco, "One Immigrant's Path from Cleaning Houses to Stanford Professor," CNN, July 22, 2016, https://money.cnn.com/2016/07/21/news/economy/chinese-immigrant-stanford-professor/.

223 **"like a God of a Go player":** Cade Metz, "Google's AlphaGo Continues Dominance with Second Win in China," *Wired,* May 25, 2017, https://www.wired.com/2017/05/googles-alphago-continues-dominance-second-win-china/.

224 **the Chinese State Council unveiled its plan:** Paul Mozur, "Made in China by 2030," *New York Times,* July 20, 2017, https://www.nytimes.com/2017/07/20/business/china-artificial-intelligence.html.

224 **One municipality had promised $6 billion:** Ibid.

225 **Fei-Fei Li unveiled what she called the Google AI China Center:** Fei-Fei Li, "Opening the Google AI China Center," The Google Blog, December 13, 2017, https://www.blog.google/around-the-globe/google-asia/google-ai-china-center/.

CHAPTER 15: BIGOTRY

229 **"Google Photos, y'all fucked up":** Jacky Alcine tweet, June 28, 2015, https://twitter.com/jackyalcine/status/615329515909156865?lang=en.

230 **In a photo of the Google Brain team taken just after the Cat Paper was published:** Gideon Lewis-Kraus, "The Great AI Awakening," *New York Times Magazine,* December 14, 2006, https://www.nytimes.com/2016/12/14/magazine/the-great-ai-awakening.html.

234 **the conference organizers changed the name to NeurIPS:** Holly Else, "AI Conference Widely Known as 'NIPS' Changes Its Controversial Acronym," *Nature,* November 19, 2018, https://www.nature.com/articles/d41586-018-07476-w.

235 **Buolamwini found that when these services read photos of lighter-skinned men:** Steve Lohr, "Facial Recognition Is Accurate, if You're a White Guy," *New York Times,* February 9, 2018, https://www.nytimes.com/2018/02/09/technology/facial-recognition-race-artificial-intelligence.html.

235 **But the darker the skin in the photo:** Ibid.

235 **Microsoft's error rate was about 21 percent:** Ibid.

235 **IBM's was 35:** Ibid.

236 **the company had started to market its face technologies to police departments:** Natasha Singer, "Amazon Is Pushing Facial Technology That a Study Says Could Be Biased," *New York Times,* January 24, 2019, https://www.nytimes.com/2019/01/24/technology/amazon-facial-technology-study.html.

236 **Then Buolamwini and Raji published a new study showing that an Amazon face service:** Ibid.

236 **the service mistook women for men 19 percent of the time:** Ibid.

236 **"The answer to anxieties over new technology is not":** Matt Wood, "Thoughts on Recent Research Paper and Associated Article on Amazon Rekognition," AWS Machine Learning Blog, January 26, 2019, https://aws.amazon.com/blogs/machine-learning/thoughts-on-recent-research-paper-and-associated-article-on-amazon-rekognition/.

237 **artificial intelligence suffered from a "sea of dudes" problem:** Jack Clark, "Artificial Intelligence Has a 'Sea of Dudes' Problem," *Bloomberg News,* June

27, 2016, https://www.bloomberg.com/professional/blog/artificial-intelligence sea-dudes-problem/.

237 **"I do absolutely believe that gender has an effect":** "On Recent Research Auditing Commercial Facial Analysis Technology," March 15, 2019, https://medium.com/@bu64dcjrytwitb8/on-recent-research-auditing-commercial-facial-analysis-technology-19148bda1832.

238 **they refuted the arguments that Matt Wood:** Ibid.

238 **They insisted that the company rethink its approach:** Ibid.

238 **they called its bluff on government regulation:** Ibid.

238 **"There are no laws or required standards":** Ibid.

238 **Their letter was signed by twenty-five artificial intelligence researchers:** Ibid.

CHAPTER 16: WEAPONIZATION

239 **inside the Clarifai offices in lower Manhattan:** Kate Conger and Cade Metz, "Tech Workers Want to Know: What Are We Building This For?" *New York Times*, October 7, 2018, https://www.nytimes.com/2018/10/07/technology/tech-workers-ask-censorship-surveillance.html.

240 **On Friday, August 11, 2017, Defense Secretary James Mattis:** Jonathan Hoffman tweet, August 11, 2017, https://twitter.com/ChiefPentSpox/status/896135891432783872/photo/4.

241 **Launched by the Defense Department four months earlier, Project Maven was an effort:** "Establishment of an Algorithmic Warfare Cross-Functional Team," Memorandum, Deputy Secretary of Defense, April 26, 2017, https://dodcio.defense.gov/Portals/0/Documents/Project%2520Maven%2520DSD%2520Memo%252020170425.pdf.

241 **It was also called the Algorithmic Warfare Cross-Functional Team:** Ibid.

241–42 **by Sergey Brin and Larry Page, both educated . . . in Montessori schools:** Nitasha Tiku, "Three Years of Misery Inside Google, the Happiest Company in Tech," *Wired,* August 13, 2019, https://www.wired.com/story/inside-google-three-years-misery-happiest-company-tech/.

242 **Schmidt had said there was "clearly a large gap":** Defense Innovation Board, Open Meeting Minutes, July 12, 2017, https://media.defense.gov/2017/Dec/18/2001857959/-1/-1/0/2017-2566-148525_MEETING%2520MINUTES_(2017-09-28-08-53-26).PDF.

244 **the Future of Life Institute released an open letter calling on the United Nations:** "An Open Letter to the United Nations Convention on Certain Conventional Weapons," Future of Life Institute, August 20, 2017, https://futureoflife.org/autonomous-weapons-open-letter-2017/.

244 **"As companies building the technologies in Artificial Intelligence and Robotics that may be repurposed":** Ibid.

247 **On the last day of the month, Meredith Whittaker:** Tiku, "Three Years of Misery Inside Google, the Happiest Company in Tech."

247 **The night of the town hall:** Ibid.

247 **the *New York Times* published a story:** Scott Shane and Daisuke Wakabayashi, "'The Business of War': Google Employees Protest Work for the Pentagon," *New York Times,* April 4, 2018, https://www.nytimes.com/2018/04/04/technology/google-letter-ceo-pentagon-project.html.

248 **a group of independent academics addressed an open letter:** "Workers Researchers in Support of Google Employees: Google Should Withdraw from Project Maven and Commit to Not Weaponizing Its Technology," International Committee for Robot Arms Control, https://www.icrac.net/open-letter-in-support-of-google-employees-and-tech-workers/.

248 **"As scholars, academics, and researchers who study, teach about, and develop information technology":** Ibid.

249 **a front-page story about the controversy:** Scott Shane, Cade Metz, and Daisuke Wakabayashi, "How a Pentagon Contract Became an Identity Crisis for Google," *New York Times,* May 30, 2018, https://www.nytimes.com/2018/05/30/technology/google-project-maven-pentagon.html.

250 **At Microsoft and Amazon, employees protested against military and surveillance contracts:** Sheera Frenkel, "Microsoft Employees Protest Work with ICE, as Tech Industry Mobilizes Over Immigration," *New York Times,* June 19, 2018, https://www.nytimes.com/2018/06/19/technology/tech-companies-immigration-border.html; "I'm an Amazon Employee. My Company Shouldn't Sell Facial Recognition Tech to Police," October 16, 2018, https://medium.com/@amazon_employee/im-an-amazon-employee-my-company-shouldn-t-sell-facial-recognition-tech-to-police-36b5fde934ac.

250 **Kent Walker took the stage at an event in Washington:** "NSCAI—Lunch keynote: AI, National Security, and the Public-Private Partnership," YouTube, https://www.youtube.com/watch?v=3OiUl1Tzj3c.

CHAPTER 17: IMPOTENCE

251 **"I really want to clear my life to make it":** Eugene Kim, "Here's the Real Reason Mark Zuckerberg Wears the Same T-Shirt Every Day," *Business Insider,* November 6, 2014, https://www.businessinsider.com/mark-zuckerberg-same-t-shirt-2014-11.

251 **when he testified before Congress in April 2018:** Vanessa Friedman, "Mark Zuckerberg's I'm Sorry Suit," *New York Times,* April 10, 2018, https://www.nytimes.com/2018/04/10/fashion/mark-zuckerberg-suit-congress.html.

251 **Some called it his "I'm sorry" suit:** Ibid.

251 **Others said his haircut:** Max Lakin, "The $300 T-Shirt Mark Zuckerberg Didn't Wear in Congress Could Hold Facebook's Future," *W* magazine,

April 12, 2018, https://www.wmagazine.com/story/mark-zuckerberg-facebook-brunello-cucinelli-t-shirt/.

252 **reported that the British start-up Cambridge Analytica:** Matthew Rosenberg, Nicholas Confessore, and Carole Cadwalladr, "How Trump Consultants Exploited the Facebook Data of Millions," *New York Times*, March 17, 2018, https://www.nytimes.com/2018/03/17/us/politics/cambridge-analytica-trump-campaign.html.

252 **Zuckerberg endured ten hours of testimony over two days:** Zach Wichter, "2 Days, 10 Hours, 600 Questions: What Happened When Mark Zuckerberg Went to Washington," *New York Times*, April 12, 2018, https://www.nytimes.com/2018/04/12/technology/mark-zuckerberg-testimony.html.

252 **He answered more than six hundred questions from nearly a hundred lawmakers:** Ibid.

252 **questioned the effect of Zuckerberg's apologies:** "Facebook: Transparency and Use of Consumer Data," April 11, 2018, House of Representatives, Committee on Energy and Commerce, Washington, D.C., https://docs.house.gov/meetings/IF/IF00/20180411/108090/HHRG-115-IF00-Transcript-20180411.pdf.

252 **"What I think we've learned now across a number of issues":** Ibid.

253 **"Today, as we sit here, 99 percent of the ISIS and Al Qaeda content":** Ibid.

256 **Pomerleau tweeted out what he called the "Fake News Challenge":** Dean Pomerleau tweet, November 29, 2016, https://twitter.com/deanpomerleau/status/803692511906635777?s=09.

256 **"I will give anyone 20:1 odds":** Ibid.

256 **"It would mean AI has reached human-level intelligence":** Cade Metz, "The Bittersweet Sweepstakes to Build an AI That Destroys Fake News," *Wired*, December 16, 2016, https://www.wired.com/2016/12/bittersweet-sweepstakes-build-ai-destroys-fake-news/.

257 **"In many cases":** Ibid.

257 **the company held a press roundtable at its corporate headquarters in Menlo Park:** Deepa Seetharaman, "Facebook Looks to Harness Artificial Intelligence to Weed Out Fake News," *Wall Street Journal*, December 1, 2016, https://www.wsj.com/articles/facebook-could-develop-artificial-intelligence-to-weed-out-fake-news-1480608004.

257 **"What's the trade-off between filtering and censorship?":** Ibid.

257 **So when Zuckerberg testified before Congress:** "Facebook: Transparency and Use of Consumer Data."

257 **after the Cambridge Analytica data leak:** Rosenberg, Confessore, and Cadwalladr, "How Trump Consultants Exploited the Facebook Data of Millions."

258 **"We've deployed new AI tools that do a better job":** Ibid.

258 **Mike Schroepfer sat down with two reporters:** Cade Metz and Mike Isaac, "Facebook's AI Whiz Now Faces the Task of Cleaning It Up. Sometimes

That Brings Him to Tears," *New York Times,* May 17, 2019, https://www.nytimes.com/2019/05/17/technology/facebook-ai-schroepfer.html.

CHAPTER 18: DEBATE

264 **wearing a forest-green fleece:** Google I/O 2018 keynote, YouTube, https://www.youtube.com/watch?v=ogfYd705cRs.

265 **Google agreed to tweak the system:** Nick Statt, "Google Now Says Controversial AI Voice Calling System Will Identify Itself to Humans," *Verge,* May 10, 2018, https://www.theverge.com/2018/5/10/17342414/google-duplex-ai-assistant-voice-calling-identify-itself-update.

265 **released the tool in various parts of the United States:** Brian Chen and Cade Metz, "Google Duplex Uses A.I. to Mimic Humans (Sometimes)," *New York Times,* May 22, 2019, https://www.nytimes.com/2019/05/22/technology/personaltech/ai-google-duplex.html.

265 **published an editorial in the *New York Times* that aimed to put Google Duplex:** Gary Marcus and Ernest Davis, "AI Is Harder Than You Think," Opinion, *New York Times,* May 18, 2018, https://www.nytimes.com/2018/05/18/opinion/artificial-intelligence-challenges.html.

265 **"Assuming the demonstration is legitimate, that's an impressive":** Ibid.

266 **"Schedule hair salon appointments?":** Ibid.

267 **he responded with a column for the *New Yorker* arguing:** Gary Marcus, "Is 'Deep Learning' a Revolution in Artificial Intelligence?," *New Yorker,* November 25, 2012, https://www.newyorker.com/news/news-desk/is-deep-learning-a-revolution-in-artificial-intelligence.

267 **"To paraphrase an old parable":** Ibid.

268 **they sold their start-up to Uber:** Mike Isaac, "Uber Bets on Artificial Intelligence with Acquisition and New Lab," *New York Times,* December 5, 2016, https://www.nytimes.com/2016/12/05/technology/uber-bets-on-artificial-intelligence-with-acquisition-and-new-lab.html.

268 **But in the fall of 2017, Marcus debated LeCun at NYU:** "Artificial Intelligence Debate—Yann LeCun vs. Gary Marcus: Does AI Need Innate Machinery?," YouTube, https://www.youtube.com/watch?v=aCCotxqxFsk.

268–69 **"If neural networks have taught us anything, it is that pure empiricism has its limits":** Ibid.

269 **"Sheer bottom-up statistics hasn't gotten us very far":** Ibid.

269 **"Learning is only possible because our ancestors":** Ibid.

270 **he published what he called a trilogy of papers critiquing:** Gary Marcus, "Deep Learning: A Critical Appraisal," 2018, https://arxiv.org/abs/1801.00631; Gary Marcus, "In Defense of Skepticism About Deep Learning," 2018, https://medium.com/@GaryMarcus/in-defense-of-skepticism-about-deep-learning

-6e8bfd5ae0f1; Gary Marcus, "Innateness, AlphaZero, and Artificial Intelligence," 2018, https://arxiv.org/abs/1801.05667.

270 **would eventually lead to a book:** Gary Marcus and Ernest Davis, *Rebooting AI: Building Artificial Intelligence We Can Trust* (New York: Pantheon, 2019).

270 **he agreed that deep learning alone could not achieve true intelligence:** "Artificial Intelligence Debate—Yann LeCun vs. Gary Marcus: Does AI Need Innate Machinery?"

272 **"Nothing as revolutionary as object recognition has happened":** Ibid.

272 **a new kind of English test for computer systems:** Rowan Zellers, Yonatan Bisk, Roy Schwartz, and Yejin Choi, "Swag: A Large-Scale Adversarial Dataset for Grounded Commonsense Inference," 2018, https://arxiv.org/abs/1808.05326.

273 **unveiled a system they called BERT:** Jacob Devlin, Ming-Wei Chang, Kenton Lee, and Kristina Toutanova, "BERT: Pre-training of Deep Bidirectional Transformers for Language Understanding," 2018, https://arxiv.org/abs/1810.04805.

274 **"These systems are still a really long way from truly understanding running prose":** Cade Metz, "Finally, a Machine That Can Finish Your Sentence," *New York Times,* November 18, 2018, https://www.nytimes.com/2018/11/18/technology/artificial-intelligence-language.html.

CHAPTER 19: AUTOMATION

280–81 **the lab unveiled a robotic arm that learned:** Andy Zeng, Shuran Song, Johnny Lee et al., "TossingBot: Learning to Throw Arbitrary Objects with Residual Physics," 2019, https://arxiv.org/abs/1903.11239.

282 **His salary for just the last six months of 2016 was $330,000:** OpenAI, form 990, 2016.

282 **And in February 2018, Musk left, too:** Eduard Gismatullin, "Elon Musk Left OpenAI to Focus on Tesla, SpaceX," *Bloomberg News,* February 16, 2019, https://www.bloomberg.com/news/articles/2019-02-17/elon-musk-left-openai-on-disagreements-about-company-pathway.

282 **"Excessive automation at Tesla was a mistake":** Elon Musk tweet, April 13, 2018, https://twitter.com/elonmusk/status/984882630947753984?s=19.

283 **Altman re-formed the lab as a for-profit company:** "OpenAI LP," OpenAI blog, March 11, 2019, https://openai.com/blog/openai-lp/.

283 **an international robotics maker called ABB organized its own contest:** Adam Satariano and Cade Metz, "A Warehouse Robot Learns to Sort Out the Tricky Stuff," *New York Times*, January 29, 2020, https://www.nytimes.com/2020/01/29/technology/warehouse-robot.html.

284 **"We were trying to find weaknesses":** Ibid.

284 **a German electronics retailer moved Abbeel's technology:** Ibid.

284–85 **"I've worked in the logistics industry":** Ibid.

285 **"If this happens fifty years from now":** Ibid.

CHAPTER 20: RELIGION

288 **Boyden had recently won the Breakthrough Prize:** "Edward Boyden Wins 2016 Breakthrough Prize in Life Sciences," *MIT News,* November 9, 2015, https://news.mit.edu/2015/edward-boyden-2016-breakthrough-prize-life-sciences-1109.

288 **some said a machine would be intelligent enough:** Herbert Simon and Allen Newell, "Heuristic Problem Solving: The Next Advance in Operations Research," *Operations Research* 6, no. 1 (January–February, 1958), p. 7.

288 **said the field would deliver machines capable:** Herbert Simon, *The Shape of Automation for Men and Management* (New York: Harper & Row, 1965).

288 **"Among researchers the topic is almost taboo":** Shane Legg, "Machine Super Intelligence," 2008, http://www.vetta.org/documents/Machine_Super_Intelligence.pdf.

290 **"AI systems today have impressive but narrow capabilities":** Ibid.

291 **the Future of Life Institute held another summit:** "Beneficial AI," conference schedule, https://futureoflife.org/bai-2017/.

291 **Musk took the stage as part of a nine-person panel:** "Superintelligence: Science or Fiction, Elon Musk and Other Great Minds," YouTube, https://www.youtube.com/watch?v=h0962biiZa4.

291 **"No," he said:** Ibid.

291 **"We are headed toward either superintelligence":** Ibid.

291 **"All of us already are cyborgs":** Ibid.

291 **"We have to solve that constraint":** Ibid.

291 **took the stage and tried to temper this talk:** "Creating Human-Level AI: How and When?," YouTube, https://www.youtube.com/watch?v=V0aXMTpZTfc.

291 **"I hear a lot of people saying a lot of things":** Ibid.

292 **he unveiled a new start-up, called Neuralink:** Rolfe Winkler, "Elon Musk Launches Neuralink to Connect Brains with Computers," *Wall Street Journal,* March 27, 2017, https://www.wsj.com/articles/elon-musk-launches-neuralink-to-connect-brains-with-computers-1490642652.

292 **as a twenty-year-old college sophomore:** Tad Friend, "Sam Altman's Manifest Destiny," *New Yorker,* October 3, 2016, https://www.newyorker.com/magazine/2016/10/10/sam-altmans-manifest-destiny.

293 **"Self-belief is immensely powerful":** Sam Altman blog, "How to Be Successful," January 24, 2019, https://blog.samaltman.com/how-to-be-successful.

294 **"As time rolls on and we get closer to something":** Steven Levy, "How Elon Musk and Y Combinator Plan to Stop Computers from Taking Over," Back-

channel, *Wired,* December 11, 2015, https://www.wired.com/2015/12/how-elon-musk-and-y-combinator-plan-to-stop-computers-from-taking-over/.

294 **"It will just be open source and usable by everyone":** Ibid.

294 **he and his researchers released a new charter for the lab:** "OpenAI Charter," OpenAI blog, https://openai.com/charter/.

295 **"OpenAI's mission is to ensure that artificial general intelligence":** Ibid.

295 **DeepMind trained a machine to play capture the flag:** Max Jaderberg, Wojciech M. Czarnecki, Iain Dunning et al., "Human-level Performance in 3D Multiplayer Games with Population-based Reinforcement Learning," *Science* 363, no. 6443 (May 31, 2019), pp. 859–865, https://science.sciencemag.org/content/364/6443/859.full?ijkey=rZC5DWj2KbwNk&keytype=ref&siteid=sci.

296 **DeepMind unveiled a system that beat the world's top professionals:** Tom Simonite, "DeepMind Beats Pros at StarCraft in Another Triumph for Bots," *Wired,* January 25, 2019, https://www.wired.com/story/deepmind-beats-pros-starcraft-another-triumph-bots/.

297 **Then OpenAI built a system that mastered Dota 2:** Tom Simonite, "OpenAI Wants to Make Ultrapowerful AI. But Not in a Bad Way," *Wired,* May 1, 2019, https://www.wired.com/story/company-wants-billions-make-ai-safe-humanity/.

300 **DeepMind announced that Google was taking over the practice:** Rory Cellan-Jones, "Google Swallows DeepMind Health," BBC, September 18, 2019, https://www.bbc.com/news/technology-49740095.

301 **Google had invested $1.2 billion:** Nate Lanxon, "Alphabet's DeepMind Takes on Billion-Dollar Debt and Loses $572 Million," *Bloomberg News,* August 7, 2019, https://www.bloomberg.com/news/articles/2019-08-07/alphabet-s-deepmind-takes-on-billion-dollar-debt-as-loss-spirals.

301 **Larry Page and Sergey Brin, DeepMind's biggest supporters, announced they were retiring:** Jack Nicas and Daisuke Wakabayashi, "Era Ends for Google as Founders Step Aside from a Pillar of Tech," *New York Times,* December 3, 2019, https://www.nytimes.com/2019/12/03/technology/google-alphabet-ceo-larry-page-sundar-pichai.html.

CHAPTER 21: X FACTOR

305 **"When I was an undergrad at King's College Cambridge":** Geoff Hinton, tweet, March 27, 2019, https://twitter.com/geoffreyhinton/status/1110962177903640582?s=19.

306 **This was the paper that helped launch the computer age:** A. M. Turing, "Article Navigation on Computable Numbers, with an Application to the Entscheidungsproblem," *Proceedings of the London Mathematical Society,* vol. s2-42, issue 1 (1937), pp. 230–265.

308 **built face recognition technology that could help track:** Paul Mozur, "One Month, 500,000 Face Scans: How China Is Using AI to Profile a Minority," *New York Times,* April 14, 2019, https://www.nytimes.com/2019/04/14/technology/china-surveillance-artificial-intelligence-racial-profiling.html.

308 **Google general counsel Kent Walker:** "NCSAI—Lunch Keynote: AI, National Security, and the Public-Private Partnership," YouTube, https://www.youtube.com/watch?v=3OiUl1Tzj3c.

309 **During his Turing lecture in Phoenix, Arizona, the following month:** "Geoffrey Hinton and Yann LeCun 2018, ACM A.M. Turing Award Lecture, 'The Deep Learning Revolution,'" YouTube, https://www.youtube.com/watch?v=VsnQf7exv5I.

309 **"There are two kinds of learning algorithms":** Ibid.

309 **"There is a wonderful *reductio ad absurdum*":** Ibid.

INDEX